SCRUDE

Mark Newham

Published by MoriartiMedia.com

First edition 2023 published by MoriartiMedia.com
Copyright © 2023 Mark Newham
Cover illustration by Mark Newham
All photographs by Mark Newham Copyright © 2023
Maps courtesy of Vidiani.com

All rights reserved. Without limiting the rights under copyright reserved above, no part of this publication may be reproduced, stored in or introduced into a retrieval system, or transmitted, in any form, or by any means (electronic, mechanical, photocopying, recording, or otherwise) without the prior written permission of the copyright owner of this book.

Except where indicated, the names of all individuals and commercial organisations with connections to incidents and events outlined in this book have been changed to protect the identities of those concerned. Any resemblance to actual persons, living or dead, or to actual commercial organisations is purely coincidental.

ISBN 978-1-7396498-2-1

About the Author

After a journalism career stretching back to the days of steam-driven typewriters, telex machines and hot metal presses **Mark Newham** now devotes his time to producing books on a variety of issues.

Published in 2011, his first book – ***Limp Pigs*** – exploded the myth of a China changing beyond recognition, became an instant Amazon best seller and was equally instantly banned in China.

The BBC called it 'Unique... Inspiring...'

He can (sometimes) be contacted through mail@moriartimedia.com

Contents

PART ONE – UK 1973 CHAPTERS 1.1 – 1.2	1
PART TWO – ETHIOPIA 1974 CHAPTERS 2.1 – 2.9	27
PART THREE – GABON 1975 CHAPTERS 3.1 – 3.9	123
PART FOUR – SAUDI ARABIA 1976 CHAPTERS 4.1 – 4.5	201
OTHER WORKS BY MARK NEWHAM	247

Part One
UK 1973

Abandon hope all who enter here

Chapter 1.1

OK. I admit it. It's true. While you were getting cold and wet and going hungry in one of the worst recessions in living memory I was not only living it up at your expense but adding fuel to the fires of that recession and contributing heavily to other problem issues of the day. Yes. I confess. That was me. I was the one taking the oil industry's grubby shilling in the early 1970s. A pay-off that made me automatically liable – in part at least – for many of the ills that befell you and the rest of the world both then and since.

So there you have it. You have your confession and I invite you to do your worst. I deserve it. What I did back then is worthy of the most severe reprisal. Not just because, having trousered that shilling, I was guilty of helping keep oil prices sky high – the basic underlying cause of the seventies recession – but because my involvement in oil industry actions also contributed significantly to the climate change problem. Had the filthy lucre I got been directed not at me but towards the development of alternatives to the burning of oil, not only would that recession not have been so crippling but we wouldn't now be facing global climate meltdown. Nor an almost exact replica of that seventies economic meltdown fifty years on.

So yes, I and others like me have a fair bit to answer for. Of that there can be no doubt. But before judgement is passed on what looks to be an open and shut case, might I be allowed a moment to draw your attention to one or two aspects of the matter which will, I hope, at least shed light on how it was that people like me got sucked into that oil world vortex in the first place? How it was that we were persuaded to assist in oil industry activities that flew in the face of one of the cardinal laws of the universe?

What it all boils down, I suppose, is this – a combination of our own desperation and the oily disingenuous tongue of an industry that was out to entangle us in its sticky deceitful web of arrogant self-interest for its own ends.

In those days it didn't take much doing. Not when, in the middle of that recession, our employment prospects were as bleak as our chances of ever clearing our overdrafts. In such a state of desperation it didn't take much to persuade us to go blind to the likely effects of our actions and sign willingly on the dotted line… even with the likes of Sir Isaac Newton looking over our shoulders and urging us to think twice before signing lest there be consequences.

'To every action there's an equal and opposite reaction!' anyone who'd not gone purposefully deaf to the warning contained in his Third Law of Motion would have heard him reminding us. 'In this case, a reaction that could put the very survival of the planet in jeopardy in years to come! The risks of bringing even more oil into the world are so great they put all else in the shade. But I think you already know that.'

He wouldn't have been wrong. Even those with dollar signs for eyes weren't blind to the enormous potential costs of our actions.

'But what choice did we have?' I can hear them appealing to the great physician. 'In the seventies it was a case of take this job or starve, almost literally. We had no option!'

'That's as may be,' I can equally hear him retorting. 'But did you really have to throw yourself into cooperating with the oil companies with such gusto? Knowing what the outcome would be, couldn't you have at least confined yourself to activities less likely to be such a direct contributor to the production of oil and all the problems it brings?'

If I might be allowed to answer that, Sir Isaac, I think you'll find it's not as simple as that. Were you to be in possession of all the facts, you might find there were circumstances over which we had no control. Circumstances in which it was a matter of gusto or nothing. As in doing anything and everything our masters required of us or find yourself out on your ear.

'OK then,' I would hope his response might be. 'Being a man of science and not one who simply accepts things at face value, I'm listening. But be in no doubt. Considering it's your utter and deliberate flouting of my Third Law that's being assessed here, it'd better be bloody good.'

Very well, then. I thank you for that, Sir Isaac, and will take up the offer to make my case in detail holding nothing back. Not that you'd probably want the whole sorry saga, that is. Some of it might have you wishing you'd been a tad more careful for that which you wish. But since I have no way of telling how much you'd find relevant and how much to be no more than the tale-telling twitterati of a travelled-out old troubadour, herewith below for your edification is the whole thing in all its unexpurgated nakedness along with the following disclaimer. Enter at your own risk and don't say I didn't warn you.

***

Let's start with a bit of history. As a man of science, Sir Isaac, I would assume you'd require such a preamble in order to set the matter in context. So…

… first it's important to remember that, in those days, things were a bit different to today. People back in the seventies were less swift to condemn those joining the hunt for new sources of oil after a war in the Middle East led to an oil embargo on the West, the price of oil trebled overnight and the world went into an economic slump to rival anything seen in the 1930s. With basic survival trumping any concerns over the likelihood of increased oil burning being the cause of further environmental destruction, no one signing up to go in search of alternatives to Middle East oil risked being tarred with the environmental pariah brush.

Quite the reverse, in fact. With the world running on empty, anyone willing to join the oil sniffing out crusade actually found themselves being elevated to the ranks of local hero, a saviour riding to the rescue of a planet crippled by global stagflation, soaring unemployment and social deprivation spiralling out of control.

As one with a background in the geosciences and some experience of field survey work it was a call to arms I could hardly ignore. Indeed, had I done so pariah status 1970s-style would have been all but assured. The white feather lay in wait for anyone with my sort of skill set claiming exemption on the grounds of flat feet or their peace time equivalent.

So, as 1973 dribbled to a close, I wrestled all environmental concerns into a box, turned the key and set off on the first of a succession of exploration assignments friends and family were convinced would involve risking life and limb in some of the most hellish conditions anywhere on the planet.

It was an image of the work I felt disinclined to disabuse them of despite being assured to the contrary by the man doing the recruiting. Left behind to suffer hours-long queues for petrol, fuel and energy rationing, a crumbling jobs market and having to skimp and save just to survive, I thought they might not appreciate hearing how I was, according to the recruiter, off to live the life of Reilly in the warmth of the tropics being rewarded handsomely for undertaking the sort of adventure most people of intrepid nature would pay to be a part of.

As a raw twenty-three year-old not long out of college and with a bank manager snapping at his heels, it was sales patter I found hard to resist. Here was the chance to travel the world at someone else's expense putting all my college learning into effect.

And I'd be solving both the world's current energy problems and my own financial ones in one hit. What was not to like?

Well, yes, there was still the little matter of an environmental conscience clamouring to be let out of the box it'd been locked away in, but that was something I could turn a deaf ear to for the moment. Far more important was getting off the unemployment cliff edge I and everyone I knew was teetering on – a predicament that wasn't even our fault. The blame, I'd been told, could allegedly be laid squarely at the door of one Henry Kissinger.

***

Before the then-US Secretary of State reportedly issued a throwaway comment at a meeting with Middle Eastern oil ministers, it looked like the economic woes facing Britain in the early 1970s were transitory and could be overcome.

But then, Kissinger is said to have remarked casually that the Middle East's economic fortunes would benefit from an upward tweak in the price of oil and the end product was a catastrophic downturn in everyone else's.

Up to that moment, although things had been difficult in the UK, it did look like there was light at the end of the tunnel, especially for people like me who'd been told time and again that with a good education the world was our oyster. Most I knew in the late-1960s had duly followed the advice of those we trusted to know best and had eagerly signed up for university courses, emerging with varying degrees of success and ready to wolf those hard-won molluscs down.

Unfortunately, by the time we were in possession of our precious oyster-netting pieces of paper in 1971 the boom of the sixties was over and it was mussels, not oysters, on the menu. The jobs market had sunk without trace and we were left with no option but to take any shellfishy thing going.

In my own case, it was a position as a lowly field surveyor with one of the UK's property development giants, a job that lasted only until things in the building industry started getting sticky and it was a case of last in first out for those of us who'd sold their souls to the world of furtively constructing ticky-tacky houses in other peoples' backyards.

When that eventuality inevitably came my way I found myself back at square one and experiencing what so many others had already discovered. That no matter how many pieces of paper one had to wave there was no one to wave it at and my own piece of paper was now clearly no more valuable than a betting slip on a horse with a fear of fences. No matter how hard I waved it no one looked and just to keep body and soul together I eventually resorted to taking any bum job going.

In just eighteen months I gained what they now call life experience through working as a hospital porter, marquee erecter, factory sweeper-upper, school dinner deliverer, laundry worker and ultimately deck chair attendant on Brighton Pier. But at least I was employed which is more than could be said for large swathes of the population suffering so badly it looked like things couldn't get any worse.

How wrong they were.

As the tail-end of 1973 approached, the Middle East war sent the price of oil skyrocketing, inflation soared and the UK plunged into the sort of recession most saw as heralding their own economic death warrants.

Most, that is, but not all. There were still those who prospered in the gloom, every one of them a devotee of the mantra of never letting a good crisis go to waste and one such was a man who'd regularly drunk me under the table at college.

To spare his blushes (though God knows why) let's call him HJ, a man who, though I'm sure he doesn't know it, was solely responsible for my entry into the oil business and who's name is still reverberating round the West African jungle after I'd plunged the expedition Landrover into a snake and scorpion-filled pit and found myself in need of someone to curse.

***

I'd only recognised HJ on doing a double-take after the man I was asking for the five pence hire of the deckchair had proffered a five pound note and told me to keep the change.

Could it be? Surely not. This tanned, besuited, embodiment of affluence could not be the tattered penny-pinching excuse for a human being I'd shared a college microscope with during geology practicals. The one bleached so white by Welsh valley mists that, on the rare removal of his tea-splattered polyester shirt, was so reflective that all around were blinded by the rebounding light.

My God! It was. And now he was looking at me smiling with teeth improved out of all proportion from the stained jagged assortment of rotting tree stumps he used to take pleasure in displaying to unfortunate onlookers just a couple of years earlier.

'How you doin' boyo?' he grinned. 'See you've made something of yourself then. Always knew you'd end up in a position of responsibility,' he smirked, indicating my ticket machine and money clip.

He'd have found his freshly minted molars coming into close contact with them if I hadn't been rendered both dumb and immobile by the sight of this human metamorphosis. As it was, all I could do was sink into the chair next to him and stare.

'Wassup boyo?' he beamed back. 'Never seen someone in from the desert before? Amazing what a few months in the Sahara can do. You should try it. Put some colour back in your cheeks it would. And cash in your pocket. Looks like you could do with a bit of both, if you don't mind me saying so.'

For the first time in the five years I'd known him this former walking cadaver had my full and undivided attention. Whatever had been responsible for this transformation, I wanted some of it and I wanted it now and he wasn't going anywhere until he spilled the beans.

It wasn't difficult prising them out of him. The Welsh windbag was full of it, more chapters and verses than I had need of pouring out of him over a full hour of ignoring all other deckchair rental wannabes at the risk of losing a job others would kill for.

Once he'd gone, off to track down the 'oasis' he was meeting someone in, I resumed my duties in such a distracted state I'd have fired myself if I'd been my boss. While I'd been scratching around trying to make ends meet in a rain-sodden, economically moribund Britain, he'd been feasting on the inexhaustible nipple of the oil industry being paid so much he could afford to tip what to me back then was the equivalent of a day's wages.

On leaving college he'd stumbled on an opening for a well logger on an oil rig and had somehow managed to persuade the interviewer he was the man for the job. OK, it was in the middle of the desert and it was in alcohol-free Libya where slaking his unquenchable thirst for beer had had to be bottled for six months at a time, but this little inconvenience had been more than worth it. Two years into a job that came complete with a canteen catering for his every non-alcoholic desire he'd amassed not only some flesh on his bones but enough beer tokens to be able to kill both himself and half the pub with him if he felt like it.

Not being able to spend anything while on site and with a healthy extra hardship allowance added on, he'd return on trips back to blighty so flush with cash he had publicans gagging for his custom, often the same ones who'd earlier barred him from their establishments. Transformed after weeks at a time under a burning sun with nothing to do but work sleep and eat, by the time they recognised who it was they were serving it was too late.

My God, how I could do with a bit of that. Were there any openings for the likes of me?

'Might be, boyo. I'll ask around when I get back and let you know.'

***

Six weeks of being ambushed and of having to give the same negative response to every beseeching question I posed and the postman had started avoiding me. Unless HJ's letter had got lost in transit from Tripoli, it was clear that beneath that transformed exterior the man was still the same unreliable communicator he'd been at college and all that talk of looking for an opening for me had been about as sincere as all his college talk of paying back what he owed.

But there was still an upside to meeting him. Maybe there was hope for me after all, I thought. If the oil industry could see something in that scrofulous dishevelled dishrag of a man surely there were openings for someone who'd actually done some work at college. Maybe not the same sort of opening – HJ, when he did go to lectures, had specialised in the rock identification discipline of petrology while I'd favoured stratigraphy – but something close. Maybe in the realm of geophysics? For one with a degree in the geosciences it was such an obvious career choice. Why hadn't I thought of it before?

Probably because what career advice we'd had at university had come solely from our lecturers, people who knew nothing about anything outside of lecturing. And so as to protect their own positions from newly-qualified graduate upstarts like us, whenever asked about routes into that profession the standard response was always the same – don't bother. Universities only took the crème de la crème so under-achieving degenerates like us would do well to look elsewhere.

Which left us where? Making a collective trek to the labour exchange and, in desperation, to the pitiful excuse for a local government-run career guidance office located above it. Required to take a career aptitude test before any benefit was paid, we duly went through the motions of ticking the test's multiple answer question boxes and waiting while the tester ran a weary jaundiced eye over them.

Called individually to her private chambers to discuss her findings it was clear it was going to be a one-sided conversation. Having clearly not listened to a word I'd said, the woman proceeded to robotically pronounce me best-suited to a career in landscape architecture and that should I so wish, an unpaid trial placement with one such a company in this field could be arranged.

She didn't have to tell me what would happen if I declined. Her eyes said it all: 'Spit on my offer and don't blame me if your application for unemployment benefit evaporates into thin air. I'll give you a week to think about it.'

Slumped round the pub table later with four other aptitude-tested classmates, the conversation inevitably turned to comparing notes on our respective 'guidance'.

First up was Bernie from Birmingham who, as his story unfolded, found the eyes of the gathering not only fixed on him but widening in stunned disbelief. Apart from one, we'd all been given the exact same 'guidance' and the same unspoken threat. Only Eddie from London's East End had escaped it. A mischief-maker at the best of times, Eddie had ignored every test question and simply ticked boxes at random. Which made him, the tester said, best suited to something that cracked even the face of this cement-faced woman.

'Groovy,' he'd responded. 'I'll get on it as soon as I get home.'

'Drinks are on you,' we said in unison. 'You're the only one whose dole they can't cut. Let's see them find an unpaid placement for you as a shepherd in Bermondsey.'

Hoisted by his own mischievous petard, Eddie had no option but to gather up the glasses we were holding out and transport them to the bar for a refill.

While he was gone the rest of us drew lots. Not knowing what agreement to work unpaid would be condemning us to we decided the best way to find out was for one of us to test the water and report back.

Dammit. It was me and a couple of days later I was reporting to the muddy, fenced-off site the said landscaping company was in the process of converting into an even muddier one.

'You the new trainee?' said the man on the gate. 'Good. Report to that hut over there where they'll kit you out.'

Kit me out? For work as a landscape architect? What did I need? A special type of pencil?

Something bigger as it turned out. Something that first left me speechless then rather less so.

Before turning on my heel and walking straight back out, the proffered 'kit' was firmly returned to the one doing the kit-giving together with a message for his supervisor. For their information, I told him, I wasn't sure I was fully qualified to use the kit provided. Not included in the curriculum of any of the three years I'd spent at university was any tutorial on how to use a shovel.

***

Before the UK economy went completely rotten, everyone in the class of '71 did eventually manage to find something in which all that book learning would not go to waste, one or two even as lecturers.

In my own case it was survey work. But when the economic wheels fell off, along with any number of others I found myself plunged straight back to square one where the only form of surveying done was that of bars and streets for accommodating female company and of newspaper job ads for anything applicable to my qualifications.

With little in the way of either presenting themselves it seemed I was destined to remain both workless and unaccommodated for the foreseeable future, a vision that eventually led to having thoughts of just giving in and signing up for the life of a Foreign Legionnaire.

The admission blurted out during my little chat with HJ had him slapping his thigh in mirth.

'In a way I suppose that's what I did,' he guffawed, 'only with rather more fringe benefits.'

OK, I told him. I'll bite. How had he dug himself out of this economic mire?

'Tip-off from a mate who sells ad space in a national paper and owed me a favour. He dealt with the company placing the well-logging job ad and gave me a nod and a wink before it appeared.'

Made sense. HJ always had been an unscrupulous bastard. But at least he wasn't a secretive unscrupulous bastard and without much prompting I was soon in possession of the whole despicable story, one that got me thinking while waiting for the letter that was never to arrive. After what he'd told me, maybe I was looking in the wrong place for the type of job I was after. That ad had been destined to appear in a section devoted to engineers, a section I never scanned assuming it to be for TV repair men, car mechanics, bridge builders and the like.

Well I would now. Obsessively. For weeks afterwards I was up before dawn to get early sight of those precious ads but if oil companies and their contractors did have jobs on offer they weren't advertising them anywhere I looked. Either that or such plum positions were going to people with far better connections than me. People like bloody HJ who was now back in the desert raking it in thanks to once having no funds for a pint and agreeing to clandestinely switch his unusually alcohol-free pee with that of a mate facing a drink driving charge.

With no such favour owed to me and jobs relevant to my qualifications not even advertised it looked like I had no choice. It'd be selling Brighton Pier deckchair tickets for me until the holiday season ended and the last, dreaded thing to do before traipsing off to the dole office, was to stack and cover the deckchairs.

When that time did eventually come it felt like we were building our own gallows, a structure I began to think might well find itself being put to that very use if nothing happened to exorcise the shadow of the Grim Reaper looming over the long bleak employmentless winter stretching out before me.

If I could scrape together the wherewithal for a rope of sufficient tensile strength and the structure didn't collapse, in the current economic climate its application towards that terminal end would be a blessed, liberating relief.

Chapter 1.2

Wandering the Brighton seafront jobless and aimless on a bleak autumnal day my eye turned from the deckchair pile to the struts of the pier. Were they more or less likely to be able to support a man's weight and the rope around his neck? They looked about as rusted and decrepit as my future prospects.

Holiday season and job now long gone my days were passed on such pensive walks when not writing yet another job-seeking letter to yet another oil company or supplementing my dole money with short-lived cash-in-hand 'contracts' with anyone with a task to do and no one to do it. Just like about everyone else I knew.

Precious few were not in the same boat and such kindred spirits found themselves gravitating to one another for warmth, company and the occasional trip to the bedroom, the one place a bit of vigorous cost-free fun could be had.

It was in one such bedroom where I received news which, although I didn't know it at the time, was to change my life forever.

The day was 17 October 1973 and I was lying in the bed of my wild, flame-haired temptress companion of the time eating toast and listening inattentively to the radio after she'd bounded enthusiastically off for a tutorial on Chinese communist poster illustration at the art college.

As punishment for supporting Israel in its recent spat with Egypt and Syria, I vaguely remember the news reader saying, the Middle East's prime oil producers had ganged up to place an embargo on oil supplies to Israel's supporters, a development that would likely lead to supply shortages in the US, the UK and across the western world.

Oo-er, I remember thinking. That doesn't sound good. The Middle East was where most of the UK's oil and gas came from and without it most of the country's road transport would grind to a halt. Surely though, the West's governments wouldn't allow that to happen. If past Arab froths-at-mouth were anything to go by some sort of appeasement agreement would be reached and pretty soon we'd be back to situation normal wondering why we'd ever got our knickers in a knot.

So nah, nothing to get fussed about and with that I rolled over and went back to sleep until it was time to traipse to the dole office to sign on.

Had I known at the time what repercussions that news would have, I might not have slept on so soundly. This time, it seemed OPEC – the group of premier oil producers no ordinary person had heard of before 1973 but which was to become a household name in succeeding weeks – was serious. Nothing the US or anyone else could say or do could persuade OPEC to pour its oil on troubled waters and before anyone knew it the embargo was affecting everyone.

In just a matter of days the price of crude oil started climbing from its pre-embargo level of around $4 a barrel and didn't stop until it eventually semi-stabilised at around $12, a price which achieved both Kissinger's alleged economic recovery 'advice' to Middle Eastern oil ministers and a rise in the cost of the West's petrol and diesel to beyond the reach of the regular motorist-in-the-street. And things didn't stop there.

With the spiralling price of oil making it more economic to rely more on coal-fired power stations than oil-burning ones the government ramped up coal-burning, a move that left the country more dependent on the mines than it had been for years and the miners saw their chance. A national coal mining strike for better pay was called and it wasn't long before power station coal stocks started running low.

Determined not to give in to the miners' demands, an intransigent Tory government under the Prime Ministership of Edward Heath fought back through an appeal to the British public to see the strike as the miners holding the country to ransom and to join the fight against it by cutting their use of electricity wherever possible.

When that failed, Heath resorted to a series of quite bizarre power-saving measures including closing TV broadcasts down at 10.30pm to prompt people to go to bed early and thus reduce home heating and lighting needs.

Feeling not a little peeved over having their bedtimes effectively controlled by a government department, the move backfired badly on Heath who was soon facing calls from all sides to settle the dispute. But he remained resolute. The government would not tolerate being held hostage, he declared in a speech he hoped would appeal to the Dunkirk spirit in the British people.

It didn't. And neither did a follow-up measure reducing the regular working week to three days to cut industrial power use. All that did was send the country into a downward economic spiral from which Heath was never to recover, losing his own seat of power a year later.

Although other parts of the western world weren't facing similar coal shortages, continuing oil supply disruption hit their economies almost as hard and as winter descended the dreaded spectre of stagflation – falling economic output coupled with out-of-control rising prices – and mass unemployment loomed ever larger.

No one looked like escaping it, least of all me and my friends who began to see any hope of getting back on our feet receding by the day. The only thing going up was prices. Everything else was heading south and the only surprise was that the struts of Brighton Pier hadn't started bending under the weight of the dangling corpses of those who just couldn't take it any more.

The saving grace, I think, was the degree of community spirit amongst our little gang. Looking out for one another became the watchword in those dark days, especially as autumn gave way to winter and the prospect of a Christmas without cheer edged ever closer. Everyone knew we'd need to be vigilant for signs of the ultimate despair descending on those worst hit by the conditions, a situation which made it even more difficult keeping from them the news I'd just received.

***

Finally caught out when one of them saw me shopping for a pair of expensive sunglasses, I found myself with no option but to fess up. The only reason people shopped for sunglasses under lardy British winter skies was because they were about to jet off to either a very cold place or a very warm one or they were about to join a blues band. And since everyone knew my guitar skills could never be described as virtuoso that left only one of the first two.

'OK, OK,' I at last admitted to the flame-haired temptress, 'I've got a job.'

A bit more arm-twisting and the news was out. All that scanning of the engineer section job ads had finally paid off. Out of nowhere an ad had appeared from an American company we'll call, to protect the few innocent parties concerned, SuperSeismic Surveys Inc or 3S which was seeking geophysical surveyors and after the most rudimentary of interviews they'd had but one question to ask – when could I start? They were under pressure from their oil company client to get a geophysical survey underway soonest and they needed me there like yesterday.

Resisting the temptation to drop everything and just disappear from sight, I tried to reassure Flame Hair, I'd managed to negotiate a short stay of execution. Time enough to at least go through all the necessary medical and financial preparations and to build up the courage to drop the bombshell on friends and family, people I'd be leaving behind to suffer the crippling effects of a global recession through what the forecasters were saying would be a particularly unforgiving winter.

Once my fiery companion had relayed that news to the gang, I relaxed. It was unanimously received with smiles and pats on the back, tinged only with a modicum of concern that I might be putting myself at risk in some of the most hostile places known to man. The image they had of the average oilman was one of rufty-tufty devil-may-care roughneck giants with the IQ of a haystack living off steaks the size of cows washed down with rot-gut moonshine illicitly brewed in war-torn Arab countries where the standard penalty for breaching alcohol bans was having your head chopped-off.

So every pat on the back for finding a way out of the economic hell gripping the country was accompanied by a worried look and an entreaty to be careful and to come back in one piece.

Touched by the concern, I duly gave them the assurance they were seeking without going into detail. Were they to be fully appraised as to how my 3S recruiter had sold me the job, I'm not sure they'd have been so supportive. While they were left freezing in the snow, penniless and hungry, I'd apparently be living it up in the warmth of the tropics, my every need catered for at the company's expense with every penny I earned going straight to the bank untouched.

It was an image of the job I was hardly disabused of during a visit to a Harley Street doctor to undergo the required pre-employment medical. After leaning over me to listen to my chest, the bulky Saville Row-suited ex-Guardsman medic finally removed the cigar from his heavily bewhiskered mouth and enquired as to my planned destination. My vaccination requirement would depend on where I was being sent.

'Eediobia,' I mumbled through the thermometer in my mouth.

'Ethiopia? Gosh. Lucky chap. Historic home of the Queen of

Sheba and her direct line descendent Emperor Haile Selassie. Stunning country, Ethiopia is. Source of the Nile, you know.'

'I thought that was Lake Victoria.'

'Ah, yes. That's the source of the White Nile. But Ethiopia is where the other Nile rises. The Blue one. Amazing scenery. Way better than the boring old White Nile. Better-looking women too. Arab influence. Fine high cheekbones and finer features altogether. Shame you're not going to Somalia though. Even better-looking ones there. But never go trusting one. Have the shirt off your back while you're still getting your trousers off. And best take a few of these,' he said reaching into his desk drawer and pulling out a jumbo-sized box of condoms. 'Don't want you to go getting repatriated with something unpleasant, do we? Your employer will never forgive me.'

By the end of the examination during which he remained intriguingly unforthcoming as to how he knew all this, I was in possession of a medical pass certificate, a date for my yellow fever jabs – which would apparently leave me feeling like shit 'but you'll get over it' – and a valuable insight into why I was suddenly employee flavour of the month amongst oil companies and their contractors.

In case I hadn't realised it, he said, the price of oil had gone up a bit recently. Result – potential oil resources that'd looked uneconomic before the rise were suddenly worth exploring in detail and the companies doing the exploring were calling all hands to the pumps.

'Even hands like you with few experienced fingers,' he said. 'So you shouldn't really be surprised to have got the call. There's plenty even less qualified than you being called up to do the dirty work. I know,' he said with a sniff, 'I have to inspect their miserable tawdry bodies every day of the week and there's precious few who get turned away. I'd have to find evidence of leprosy or hepatitis or something equally unsocial for that to happen.

'There's even been the odd out-and-out psychopath given the all clear, but only after I've found out where they're being posted. If it's somewhere dry they usually get the go-ahead. Alcohol and psychosis don't mix so I reckon you'll be OK. Damn good beer in Ethiopia if memory serves so you're unlikely to come across any raving nutters on your crew.'

Well that's a relief I thought as I hitch-hiked home on finding the train drivers had gone on strike. No psychos where I was going. Just raging alcoholics and sex maniacs.

***

Which category of these, I wondered during the trip, would HJ have been placed in? He qualified on almost all counts. Which was probably why he'd ended up in Libya. Get out of line there and he'd be counting himself lucky if he was repatriated short of only a hand or two.

One thing was for sure, though. I wouldn't be running into him where I was bound. Not unless Ethiopia had converted to Islam and placed a ban on alcohol without telling me. Time, perhaps, to do some digging. Just to be on the safe side.

In the brief time allowed me before being shipped out to somewhere I'd be calling home for the next two years, I started by finding out what the primary religion was and where the hell Ethiopia was anyway. School geography and history lessons hadn't dwelled much on out-of-the-way places Britain hadn't been able to colonise and occupy for an indecent length of time.

A quick visit to the reference library while waiting for my inoculation dates to come round produced enough basic information to satisfy immediate needs. Once called Abyssinia, Ethiopia was a country of ancient antecedents located in what was known as the Horn of Africa in the dark continent's north-east quarter. Almost all Ethiopians were Coptic Christian and the only time it had suffered colonisation was when Italy's dictator Mussolini muscled his way in in 1935. That lasted only until the Allies kicked him out during WW2 and Britain helped Haile Selassie back to power, a seat he still occupied as Emperor at the time I was heading there.

Apart from that I learned little except that both high and low altitude extremes were to be found here, Ethiopia was endowed with both mountains and deserts, there was swingeing poverty, its calendar was seven years behind ours and each of its years was split into thirteen months (of sunshine apparently, according to the Ministry of Tourism).

Well that didn't sound too bad. At least being Christian I'd be able to get a beer. How though, I thought? I didn't have any Ethiopian money. Might my bank be able to help?

'Where?' said the teller. 'Never heard of it.'

After a considerable delay during which she shuffled off to update her geographical knowledge she returned to inform me that Ethiopian money was called the Birr, that the bank didn't have any and that it might be a while before they got some.

'So why not take some travellers cheques and US dollars instead, dear? You can change them for local currency when you're there.'

While they were being arranged the manager called me in. Since the company I'd be working for was American and I'd be being paid from the USA, had I thought of opening an offshore account? Others like me did it to avoid paying UK tax and it was, apparently, all quite legal. He could arrange it if I liked.

Crikey. Talk about a steep learning curve. From being on the dole and dreading the arrival of any communication from the bank I was suddenly being introduced to the world of high finance wheeler dealering and being given insider financial advice from a man who'd earlier bombarded me with 'when can we expect clearance of your overdraft?' letters. It was all a bit much to take in in one go and I staggered out of the bank with my head spinning, the proud owner of not just one account but three – my UK sterling account, an offshore sterling account and an offshore US dollar account. Oh, and a credit card with a quite decent credit limit on it.

Blimey. So this is what it was like to be one of the chosen ones, I thought. But I'd have to unpuff my chest before meeting up with Flame Hair. Having already reacted with pointed thin-lipped silence and a glance at the Che Guevara poster on her wall on being told I was selling my soul to the American oil industry devil as she was preparing to go off on a Stop The Vietnam War march, I wasn't sure she'd fully approve.

Even less so if she found out my parting Tubular Bells album gift to her was bought with a capitalist running dog credit card. If that ever emerged I'd be leaving the country wearing it.

As it was, my 'chosen one' status was only amplified when she unwrapped the album with a look of undiluted delight and a vow to try to stay faithful to me until my return. More than I could ever have hoped for in an age of female independence-asserting burning of bras and a vow I wasn't sure I'd be able to match if called upon to do so.

In fact I knew I wouldn't. Not without crossing my fingers anyway. If what the doc had told me proved accurate I knew the temptations likely to be encountered in Ethiopia would be too much for a libido going into unfulfilled overdrive in the exclusive company of men in the desert.

Bless her cotton socks. As if she knew that all such a return vow request would generate would be a shuffling of feet and a swift change of subject, none was forthcoming and as we parted for the final time at the station, I breathed out. I was off on the adventure of a lifetime with no conditions attached, a pocket full of condoms and someone waiting patiently at home for me.

If I ever made it home. During a post check-in drink in the airport bar with a 3S colleague going back after his annual leave, the talk turned to how friendly the natives were and the news wasn't good.

By and large they were fine, said Danny Boy, my wild-eyed Irish counterpart attached to one of 3S's other crews. 'But beware the Afar. They've got the reputation of being one of the most savage warlike tribes in Africa. Slit your throat without a second thought if they think you're giving them a sideways look. Then they'll rip your teeth out and wear them on a trophy string round their neck. Don't have any dentures fitted do you? They're especially highly prized.'

Shee-it. I did.

Part Two
ETHIOPIA 1974
Nothing ventured nothing gained

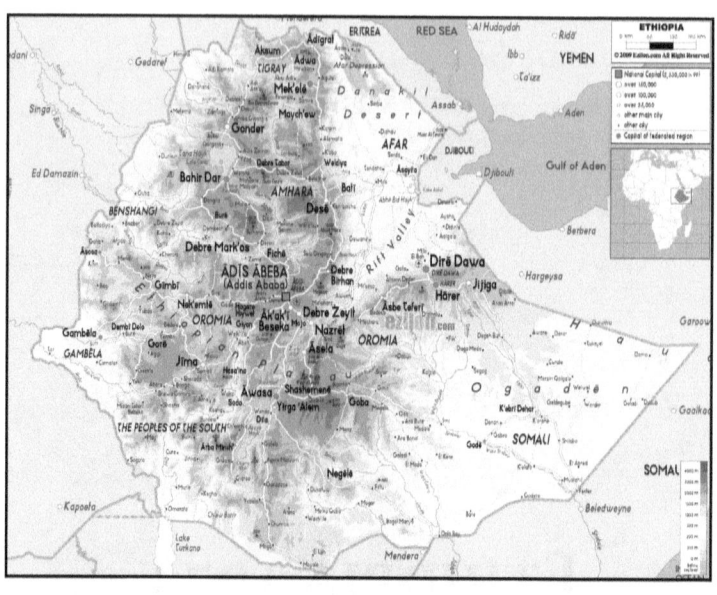

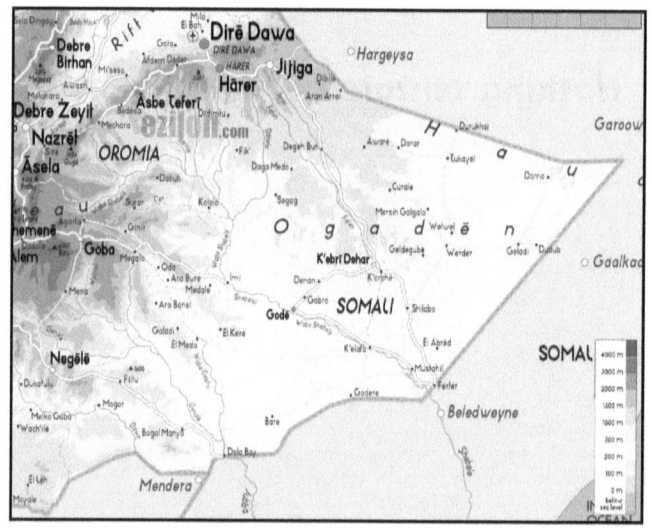

Chapter 2.1

Twelve hours of like hair-raising illuminations into what I'd let myself in for and we arrived in the Ethiopian capital Addis Ababa with Danny Boy still talking and me knowing more about the vicissitudes of the locals, the ferocity of the wildlife and the deadly unpredictability of the weather than was comfortable for a man on his first proper trip outside home turf. The only time he'd shut up was to hear my excuse for almost leaving the crew's mail behind but all that did was give him the opportunity to launch into yet another cautionary tale.

'Last time a new boy arrived without the mail,' he said, handing me the packet I'd left at check-in, 'he was stripped naked, taken out into the desert and left to find his own way back. So reckon you owe me a beer or two for saving your bacon there. And not the piss you were drinking in the bar. What was it? Shandy?'

Not a little abashed, all I could do was nod. Doctor's orders, I admitted on being pressed. No strong drink until I'd finished the course of antibiotics he'd prescribed.

'Antibiotics is it! You gotta dose then?'

'Nooo. Just a touch of urethritis. Just a mild infection you don't even have to have sex to get, so the doc says. Even monks get it apparently.'

'Well jayzusmaryjoseph,' he guffawed, slapping his thigh in mirth. 'Dat's a first. People normally come BACK with the clap, not go out with it!'

'It's NOT the bloody clap. It's urethritis!'

'Call it what you like, me bucko. But I know what the rest of your crew will call it. Can't keep a ting like dat quiet for long. Someone's bound to ask when dey see you not drinking. 'Spose you could tell 'em you're a Muslim but I wouldn't advise it.'

It wasn't the only thing he advised. No longer the rookie on the crews with me now in that role, he took full advantage of his newly-acquired old hand status to give me the 'benefit' of his experience all the way to Frankfurt. Then while we waited for our connection. Then for the full duration of the bone-shaking flight to Addis in a decrepit Ethiopian Airways Boeing 707 on which the only choice of food was injera wat, an Ethiopian staple consisting of a fiery assortment of tapas-like dishes scooped up with what looked and tasted like something left over from a tyre factory.

By the time we got there my ears hurt, my stomach was in full rebellion and all I wanted to do was get to the hotel and crash out. Alone.

Fat chance. Before that we had to negotiate an airport as ragged and dilapidated as the hordes massing round the exit ready to take advantage of any greenhorn arrival. And not having entry visas didn't help, something our employer had helpfully failed to warn us about.

Well, me anyway. Danny Boy seemed fully prepared for being stopped at immigration and quizzed over the visa deficiency. He'd handed over his passport with a $10 bill in it and suggested I do the same to be granted a tourist visa for a month. It worked, and as we made our way to the chaotic baggage reclaim area he reluctantly let me in on how things worked here.

'Take a look at your ticket,' he said. 'See that? It's for an onward connection to Djibouti. So, as far as the immigration here is concerned, we're just passing through, stopping off in Addis to see the sights.'

'But we're not. We'll be working.'

'They don't know that. And they're so disorganised they're

unlikely to check. It's standard practise for companies working here. Far too much like hard work trying to arrange work visas. If they did that for every foreign worker here nothing would ever get done.'

'But what about on the way out?' Having only ever been on a plane once before on a quick trip to the Costa Brava in the late sixties I had little experience of international air travel. But surely someone would ask difficult questions when it came to our leaving long after our tourist visas had expired.

'That's for then, not now. A bit of baksheesh generally shuts them up so always best to keep a few US dollars up your sleeve for times like that.'

Of all the tips Danny handed down during the trip, this one was out on its own. If there was one thing Africa was, I was to discover in succeeding years, it was amenability incarnate. Almost any obstacle could be made to disappear with the surreptitious appearance of a dollar bill or two, the reason dollars had come to be known the length and breadth of Africa as greenback grease.

But as magical as the grease was, there were still one or two things it couldn't achieve. Things like locating my baggage.

After an hour of sweating and gasping for breath in a tropical airport baggage reclaim area with no air conditioning some 7,000 feet above sea level, we gave up. Everything had arrived except my main bag, the one with hot climate clothes in it, and no amount of greasing the wheels could magic it up.

So, with a sigh, I duly filled in the grubby lost baggage report form, took a carbon copy and headed for the taxi rank to find Danny Boy beaming beside our ride into town – the biggest, flashiest car I'd ever seen.

Since everything from here on in was covered by the company, he grinned, it seemed a shame not to take advantage of it. Likewise with our accommodation, the second best hotel in the city after the Hilton but only because the Hilton was full.

Blimey Newham, I said to myself as we swept through the endless ranks of crumbling shanties lining the road on the way to the luxurious Ghion Hotel, looks like you really have landed on your feet this time old son. If this is the standard level of treatment doled out by 3S to its employees, maybe all that interview talk of its workers being afforded the best that money can buy wasn't just so much hot air sales patter after all.

This I could definitely live with… even if it did mean going numb to the jab of red hot pokers tormenting my conscience while gliding in air conditioned comfort past the tattered, emaciated, grim-faced millions wobbling along on wonky-wheeled donkey carts under a blistering sun. If we found what we'd been sent to find, I tried to reassure myself, the proceeds would surely trickle down to them in the long run… wouldn't they?

With the ether going suspiciously unresponsive on the matter, despite suspecting the worst I finally decided to consign the question to the outcome pending file. What possible conclusion could I reach without access to some pretty serious empirical evidence, was the pathetic apology for an excuse I consoled myself with on sinking into the depths of a huge enveloping bed to sleep off the effects of the past twenty-four hours.

***

Dragging myself out of bed the next morning and running on the spot to get the blood pumping again, I instantly regretted it. Acclimatising to altitude had to be done slowly, I vaguely remembered reading somewhere as stars circled my head and I found myself collapsing onto a chair to catch my breath.

'Wow! They weren't joking,' I muttered to no one in particular. 'Hope to hell this acclimatisation thing doesn't take long or I'll be about as much use as that hay fever-afflicted surveyor I once worked with who couldn't go out in the field when the flowers were in bloom.'

'Don't worry,' Danny Boy reassured me over a sumptuous Ghion Hotel breakfast. 'Where we're going it's not altitude sickness you have to worry about, more heat stroke,' he said, a reassurance that gave him the chance to launch seamlessly into a resumption of the interminable catalogue of horror stories I'd eventually stopped listening to on the plane.

After a further fifteen minutes of them I decided it was time to assert myself and change the subject.

'Heat stroke is definitely what I'll be suffering if I don't get a change of clothes pronto,' I interjected. With my lighter-weight stuff all in the lost bag all I had was the winter clothes I'd travelled down in and they were beginning to smell.

'Think I'll have to take a trip to the shops,' I said. 'Know anywhere decent?'

'Sure. There's some in the hotel lobby. Just put them on the bill and claim the whole lot back from the company.'

'Claim it back!' I gasped. 'Hang on a minute. You mean WE have to pay for all this upfront?' I wasn't sure I had enough travellers cheques to cover it.

' 'Course. They didn't tell you that back in the UK?'

No they bloody didn't. All I'd got when popping by 3S's London office en route to Heathrow was the air ticket and the packet of mail for the crew I'd left lying on the desk at check-in during a manic, adrenalin-fuelled rush to make it to the plane on time. Somehow they'd considered a full pre-trip briefing surplus to the requirements of someone destined to travel to a place few had ever heard of. Danny Boy would fill me in on what to expect, the big cheese had muttered into the chaos of papers cluttering his desk as he'd distractedly waved me out of his office, out of his life and on my ill-prepared way.

Well, Danny Boy had done that all right. He'd filled me in on everything I needed to know... except the most important stuff. Apart from failing to warn me about the expense-claiming thing, there'd been not a word about getting only one night in Addis to recover. Something that left me no time to even browse the lobby shops for cool clothing.

Urging me to hurry up and finish my breakfast, he disappeared to pack with the news that we had to be at the Pipeyard in an hour.

'The what?'

'The Pipeyard. The airstrip our flight goes from. C'mon.'

***

Four hours later I was being disgorged at a remote desert airfield from an antiquated twin-prop DC3 in which I'd nearly frozen to death despite the winter clothing I hadn't had time to get replacements for at the Ghion.

Racing to get to the Pipeyard in time, all I'd been able to do was pay the hotel bill, count what remained of my seriously depleted stock of travellers cheques and throw myself into the car Danny Boy had impatiently revving its engine outside. Until my bag was recovered, I'd just have to make do with the clothes I stood up in.

By the time I arrived at the outback survey camp caked in bull dust from an onward trip from base camp on the back of an open-backed lorry, I didn't need to be in the clothes for them to stand up. Coupled with a three-hour flight in sub-zero temperatures in an unpressurised plane that came complete with paratrooper canvas benches running the length of the plane with parachute attachment rails ranged above then being hit in the face by a scorching desert wind on disembarking, they could do it on their own.

Shame they couldn't also walk on their own. If they had, I might have avoided the sniggers of derision received on arrival at the base camp airfield. Stiff with cold and exhaustion I'd had to hang on to the disembarkation steps handrail to prevent a headlong plunge to the concrete pad below and being dressed in warm jacket and full-length cord trousers with shoes and socks didn't help. At the sight of the over-dressed ashen-faced cripple edging his way step by careful step down the gangway clutching a carrier bag, it was all my welcoming committee could do to stop themselves pointing with quivering fingers and slapping one another's backs in fits of uncontrollable mirth.

As it was, one or two had failed to control the cowboy yelps they were trying to keep in and I swear I heard one ask another if it was now company policy to employ the terminally ill to improve their corporate image. Especially ones so poor they couldn't even afford a suitcase.

Actually, I could have been mistaken there. For all I know they were being sympathetic to my plight, so full-on was the deep south drawl my exclusively American welcoming committee used to communicate between themselves.

'Shay-eet,' said the one with the biggest Stetson and the pointiest cowboy boots, looking at me as if I was something those boots had just stepped in. 'Ah heard they wuz sendin' us a Limey but ah kinda hoped they'd be sendin' a live one. So what's your name then, boy? Lazarus?'

When I corrected him, all he could do was rub his four-day growth of stubble with a gnarled bearlike paw and inform me that that was unlikely to stick. Once my crew saw what he was now looking at they'd undoubtedly be re-christening me with some more apposite name.

'Jus like ah wuz,' he said proudly, baring a luminously white set of perfect dentures and thrusting out his paw to take mine with a grip of forged steel.

'Real nayerm's McGrew. Chayerf driller round here. But folks call me Dawg. Short fer Smells Like Dawg cuz if they's anythin' worth drillin' ah kin sniff it out. Ain't that raght, boys?'

'Yeah… er, raght Dawg. For sure. Got it in one there,' mumbled another cowpoke lookalike standing nearby, looking down at the ground and shuffling his feet as he said it.

'Anyhoo. Welcome to Gard's own asshole boy. Real nayerm

Kebri Dehar and the base for our operations in these parts. You's assigned to Crew 3 so ah'm told an' ah'll be comin' to visit one o' these fine days. But not today. First things first. Gotta package to pick up from that Irish motherfucka y'all came down with. Where is that bog-trottin' sonofabitch?'

'Right here, Dawg,' said Danny Boy from behind him. 'Got your single malt right here... and it ain't the only thing being imported,' I heard him announce with glee as the pair stalked shoulder-to-shoulder off out of earshot towards the airfield cabins. 'You'll never guess what the rookie has brought with him!'

Minutes later I was in one of the cabins myself being introduced to 'the man', a five-foot nothing tub of lard slumped behind a massive desk who only got out of his chair to reach out for the mail packet I was holding.

'Noo boy, huh?' he grunted, slumping back down and sifting through the mail without looking at me. 'OK. Need you on that truck out there in five minutes. Questions?'

'Well, just the one,' I said holding out the lost baggage report form from Addis airport. 'Is there someone who can pick up my bag and send it on when it arrives?'

Keeping his hands firmly planted on the desk, the 'man' ignored the form and returned my inquiring look with one of pure, undiluted contempt.

'What you say, boy? YOU want ME to be YOUR bellboy?' he snarled. 'Know who the fuck you're talkin' to, boy?'

'Well no, not really. We haven't been introduced and you haven't had the good manners to tell me who the fuck you are,' I thought but didn't say. His demeanour suggested now might not be the time to take issue with anything he said or to inform him I hadn't been known as 'boy' since that day in the bath when, to my horror, I discovered hairs sprouting round what my mother called my little pee shooter. So instead, I adopted the most humble posture I could muster under the circumstances and mumbled something apologetic to the floor.

The tactic worked and after a moment or two of barely-restrained seething, he simply waved me dismissively out of his office with a command to bother one of his underlings with my baggage reclaim request and to take careful note of the sign nailed above his office door.

'Office of Shit-Giver in Chief', it said, a title I'd wrongly assumed to be a reference to the unsurpassable amount of caring for my welfare I could expect from the occupant within. That, I was now in no doubt, was the polar opposite of the meaning intended. Oops.

After backing shell-shocked out of his door and standing uncertainly outside trying to interpret the sign's message I sensed a presence at my shoulder.

' 'Spect ya could do with a pick-me-up after that, mate,' the unmistakeable twang of an Australian voice was commenting. 'Quite a piece of work is our Blott,' said the archetypal Aussie ocker who'd taken me under his wing and started leading me towards the canteen trailer.

'Real name Bastides but no one calls him that after someone told him to his face he was a blot on the landscape. One of your countrymen, as it happens. Another Pom who lasted about a week. Hope you're a bit more thick-skinned than he was. Shandy is it?' he asked with a smirk as we made our way to the canteen counter.

Bloody Danny Boy! That could only be his doing. Hadn't taken him long to blab about my little health issue and subsequent drinking preference and now the news was all over camp.

Bastard. I'd get him for that next time our paths crossed, I thought, handing over the lost baggage and hotel bill claim forms to the ocker who'd said he'd take care of them for me.

***

In fact I'd double get him on learning the nickname the news had prompted my fellow crew members to saddle me with before I'd even reached the survey camp out in the scrub.

At the time I was too shattered to argue having just spent the better part of three hours being bounced around on the back of an uncovered truck as it ploughed through thick bull dust to the camp under a sun that was in no mood to take prisoners. All I wanted to do on arrival was get to the air conditioned trailer I'd been promised by the company recruiter in London, flush the layers of bull dust out of every crevice in the trailer's shower and sleep for a week.

Blinded by the dust, it took a moment or two to realise we'd finally arrived. But not at the main camp I assumed on scanning the ragtag assortment of tents interspersed with the remains of various clapped-out vehicles, strewn engine parts and assorted detritus. The driver presumably wanted to drop a couple of parts off at what must be the camp 'garage' before going on to the crew's living quarters to deliver both me and the food supplies he was carrying.

'Hang that box on me while ya there, will ya?' said a voice from the still settling dust cloud. 'No, not that one. That... there.'

Expecting the one indicated to be heavy with metal bits and pieces I braced myself to lift it, only to find it was far lighter than it looked. Strange. That didn't feel like any part of any engine I'd ever worked on.

It wasn't. It was full of packets of crisps, duly identified when the man behind the voice, Canadian by the sound of it, tore the box open and started digging through it for the right flavour.

' 'Spect ya could do with a lube or two after that trip,' said the man spraying a cloud of crisp fragments in my face. 'Follow me and I'll sort you out. We'll start a tab for you later.'

Ooookay. So lube was camp talk for a drink and the truck was obviously going to be here for a while offloading garage supplies. Not ideal but there was nothing to be done and I duly clambered down to trail behind him to the nearest tent.

Inside, slumped around a long dirty plastic table were some of the grubbiest, most dishevelled shaving soap-deprived individuals I'd ever set eyes on outside the backstreets of Brighton, haunt of choice of the town's growing retinue of homeless itinerants.

As my eyes adjusted to the gloom crisp-spray man handed me a bottle of lukewarm Fanta, not my drink of choice but something that'd at least clear my throat of bull dust.

It was vile and my face must have registered it because at that point crisp man spoke up with an apology over there being no lemonade for a shandy.

'That's your drink ain't it, Two Brains?' he said trying to keep a straight face.

Damn bloody Danny Boy. He or someone else at base camp must have radioed ahead with the news. He'll bloody regret that, I thought, scanning the faces round the table, every one of them biting their lip to prevent descents into fall-off-chair, finger-pointing hilarity.

OK, OK, I get it. Very funny. Ha bloody ha, my face said while simultaneously wondering what was with the Two Brains reference. Was he referring to me?

He was.

'Well, from what we hear that's how many you've got. One in your head, one in your dick. Neither of which seems to be in full working order after bringing a dose TO Africa while NOT bringing our mail.'

You're a bloody dead man, Danny Boy.

But that could wait. PLEASE, I begged. After the trip I'd just had could someone PLEASE just show me to my trailer before I collapsed on the floor?

At the mention of trailers, the gathering first stopped grinning then blinked uncomprehendingly at one another before bursting into snorts of thigh-slapping laughter.

'Trailer?' crisp man managed eventually. 'There's your 'trailer',

Two Brains,' he chortled pointing towards a stained, tattered, lopsided ridge tent with no door flaps or mosquito net pitched close to a leaking drum of diesel.

'All mod cons and all the comforts of home,' he grinned. 'But don't get too comfy. Camp move at first light.'

Chapter 2.2

Despite the accommodation bombshell and my desperation to get my head down regardless of what it was being put down on, I had to know.

'How?' seemed a reasonable enough response to the camp move announcement and a glance around the shambolic arrangement of dilapidated tents interspersed with broken down vehicles and strewn engine parts the occupants had the gall to call a camp.

'Ermay gawnna hay-erlp,' said the table occupant I was later to learn was the crew's semi-literate seven-fingered Louisiana swamp-born driller who was about as at home in the desert as the average duck.

My turn to blink uncomprehendingly.

'Digit says the army's gonna help us,' translated crisp man helpfully.

'Yaw,' said Digit. 'Bay-out tha owernly use they er. Ray-ert Footsie?'

'Rahght. Sure wouldn't want to have to rely on 'em to defend anything worth stealin'. Speshly ma compooter truck,' mumbled the crisp-munching Footsie, the familiar that close colleagues used instead of Big Foot, a camp nickname conferred on him courtesy of being the man at the hub of this whole seismic operation. Without him, it was eventually explained to me, the crew's monster vibroseis machines with their huge vibrating 'feet' that sent shock waves down into the rocks would have been just worthless pieces of noise-making junk. It needed someone like Big Foot to set the strings of surface-mounted geophones along the geophysical line and ensure the sonic bounce back signals they picked up got recorded on the computer in Big Foot's air-conditioned truck for conversion into pictures of the different layers of substrata beneath the line.

'Army? Computer truck?' I managed after doing a quick translation of my own.

'Yo,' said Big Foot. 'Bout as useless as each other. But at least the truck sort of works. Which is more than can be said of most of the wheels round here.'

'You gotta a complaint,' interjected the one table occupant whose level of dishevelled grubbiness made all the others look positively scrubbed, 'take it up with management, Mr Big Shot Geophysicist. I can only work with what I'm given and that's about as much as they pay the locals round here. Gimme a decent workshop and a parts budget and I can bring a buried Bedford back to life!'

'Message received,' said Big Foot calmly. 'No need to go gettin' yo fan belt in a twist, Whacko. We all know there's a mechanical magician lurking under that grime. But ya gotta admit this crew's got more cratered vehicles than anyone else. What I wanna know is how the other crews keep their fleets serviceable when ours is more scrapyard material than forecourt.'

Even under the layers of grease you could see the veins in chief mechanic Whackamole's temples beginning to swell and it was only the intervention of another crew member ducking opportunely into the tent that prevented Big Foot needing to extract a spanner from his nostrils.

'Mail's here!' announced the man dumping the packet I'd almost left at Heathrow on the table.

In one fell swoop all thoughts of escalating the difference of opinion between geophysicist and mechanic were forgotten as the entourage tore into the packet and began disappearing in all directions with their share of the spoils. In the end it was just me and Big Foot left at the table.

'No mail for you?' I enquired.

'Nah. Don't really expect any. Got family with me in Addis and my folks back in Calgary ain't the best of correspondents.'

'How often do you see them?'

'Once a year at annual leave.'

'Don't they like to see you more often?'

'Sometimes. Like it or not, they get to see me in Addis every fourth week,' said Big Foot confusingly. 'Like the rest of the crew I'm on a three weeks on, one off rota. Can be a bit rough at times getting through the on weeks but not half as rough as it is for the army. They only get a break every six months.'

Ah, yes. The army. In all the commotion I'd almost forgotten about them. Was 'army' the nickname for the local labour force Danny Boy had told me was attached to every crew?

'Nope. Got ourselves a real army. Camp's over there,' said Big Foot, pointing beyond our camp to a barely visible speck of tents away in the distance.

'Er… why? What's the army doing here?'

'Providing protection.'

'To who?'

'Us.'

'From?'

'Somali army.'

Two more tepid drinks from a tepid fridge later and all was revealed. Without having had the decency to tell me, I'd been dispatched by 3S to carry out a mapping expedition in one of the most disputed areas in the whole of the Horn of Africa – a slice of semi-arid Ethiopian wasteland known as the Ogaden desert that was long-claimed by Somalia and whose ruling clique was in the habit of ordering the occasional military incursion into it. And just in case they felt like trying their luck while we were in the vicinity – not a million miles from the Somalia/Ethiopia border – the Ethiopian government had seen fit to order a detachment of its own into the area to act as a deterrent.

Was it working?

'So far,' said Big Foot. 'Which is just as well cos the guys they've given us couldn't hit a barn at ten paces. Not the door, the barn. Had a shooting competition with 'em a couple of weeks ago and we beat 'em to pulp. Drunk as skunks on tej the whole lot of 'em. It was a miracle no one got hit.'

'And if they had been?'

'Then I guess we'd have to call in The Gravedigger.'

***

Carrying such a reassuring thought with me to my 'quarters' that night, sleep did not come easily and it was almost with relief that I was roused before daybreak for the camp move.

For sleep read backbreaking sweat-drenched few hours of tossing and turning on a stained sagging canvas cot in a tent that came complete with a squadron of local mosquitoes keen to make the acquaintance of any fresh meat delivered into their midst. And for rousing read being tipped unceremoniously out of bed by a grinning Somali houseboy ordered by the camp management to strike my tent at dawn regardless of whether the jet-lagged corpse in residence was still groaning in pain under the grubby sheet pulled up over its head.

Breakfast? Dream on. With the mess tent long gone by the time I crawled out from under my own semi-demolished lump of canvas still wearing the clothes I'd travelled out in, breakfast for me consisted of another bottle of sun-scorched Fanta and the nob end of a stale baguette that even Digit had turned his nose up at.

Result: although I didn't think it possible, by the time everything was re-erected in the new location about twenty clicks – hairy-arsed rufty-tufty oilman speak for kilometres – up the geophysical line, I was even closer to an early demise than the day before and after the most rudimentary of 'dinners' of canned corned beef and cold baked beans I fled to collapse on my sagging cot to reflect on the events of what had proved to be a somewhat eye-opening day.

With one exception, I thought, the camp move process wasn't that much different to my days working as a marquee erector for a bunch of racist slave-driving bigots in the heart of the Sussex countryside. That exception being that the marquee-transporting lorry we used to carry often still sodden swathes of canvas from site to site back then actually worked.

Not here. Not one of the crew's three British army surplus Bedford five-tonners came complete with a working engine, every one having to be hauled fully-loaded cross-country by our Ethiopian army 'protectors' with the crew's labour force clinging precariously to each Bedford's mountainous load of tents, furniture, cooking equipment, provisions and engine parts in various states of disrepair.

Had the progress of this creeping caravan of charabanc decrepitude been subject to any Somali military force monitoring, it crossed my mind they'd undoubtedly be ruing not yet having received orders to invade. Our protection 'force' might have had difficulty carrying out their primary function at the same time as struggling to help those under their protection cross steep-walled dry river beds that'd instantly turn to torrents if the rains arrived early and with no warning.

The phrase 'sitting ducks' came easily to mind in the circumstances, a phrase – or something very close to it – that was also on the mind of my fellow crew members. It was also the nickname – almost – of the man now being collectively cursed for selecting the crew's new campsite and then disappearing off site leaving the preparation job unfinished.

'Sitting Schmuck really lived up to his name on this one,' grunted Big Foot in answer to my query about the man. 'Everything the schmuck touches turns to shit and there's nothing he likes more than to leave everyone else right in it. This time the lazy drunken fuck shipped out soon as he heard you were coming to replace him leaving us to finish what he started.'

'... der mudderfuckin' asshole sonafabitch,' added Digit enlighteningly.

Scanning the new campsite on which we were supervising the tent erection by our Somali labour force I couldn't see it was much different to the one we'd just left and told Big Foot so. What was unfinished about it?

'No fuckin' airstrip.'

'Airstrip?'

'Yeah. No fuckin' airstrip, no fuckin' mail or supplies.'

'Thought all that came by truck. I bloody did.'

'Initiation,' chipped in mechanic-in-chief Whackamole with uncharacteristic grasp of the English language. Unlike the bulk of the crew to whom anything more than two syllables was usually one too many, Whackamole turned out to be not as American as I'd initially taken him to be, slipping back into his native Glaswegian when occasion demanded it. 'Oor lairds 'n masters at base camp do that tae all noo meat, laddie. Specially the Baron.'

'Baron?'

'The Red fuckin' Baron, mon. Ye did nae ken him? Oor brillyant bleddy company's head fuckin' pilot who makes sure all new arrivals get to become fully acquainted with the pleasures of ridin' the bushwhacker express. A sorta incentive so's we cut decent strips for him to land on. Shit strip, shit service. Seen the miserable sonofabitch unload half the stuff we ordered then take off on a shit strip leaving guys on the ground who were ready to go on leave. No one wants to go in or out by truck. You've seen why.'

I surely had and now I knew the background it was a lesson well-learned. If I'd translated Whackamole's guttural Glaswegian correctly, rule one in choosing and setting up a new camp – mostly the responsibility of the line surveyor, i.e. me, it seemed – was keeping the Baron happy, even if it did mean annoying the cat skinners, crew talk for the Ethiopian Caterpillar bulldozer drivers who were responsible for driving the line forward through the scrub.

Paid by the kilometre of line cut, Whackamole explained, they sometimes had to be bribed or blackmailed into agreeing to turn round and go back to cut and grade a new airstrip. The offer of a hooker or two waiting for them next time they went on leave usually did the trick, he said. Or if they were still reluctant to cooperate, the threat of getting the base camp doctor to put the word out they had the clap.

'He wouldn't do that, would he?' I fired back at the mechanic. 'It'd be counter to medical ethics.'

'ETHICS!' Whackamole exploded. 'To The Gravedigger that's a county in the east of England. The bloodless bastard'll do anything to keep the nice little number he's got going here. Nice fat paycheck at the end of each month for sitting on his fat air-conditioned arse at base camp, keeping everyone on the job regardless of whether they're actually sick or not and for helping his lord and master Blott keep Blott's little sideline enterprise going.

'When you're here a bit longer, Two Brains, you'll discover why getting a plane to deliver supplies can be as difficult as gettin' the truth oota a Sassenach. Plane only comes when Blott isn't using it to ferry hookers from town to town. Even clap-ridden ones complete with clean bills of health from The Gravedigger. Nice little earner that merry-go-round is for the pair of 'em. So nice in fact it's sometimes meant us having to threaten to withhold our labour unless they shelve their HookAir operation for a day and give priority to getting supplies out to us in the scrub.'

This, as it turned out, was not the rare event I'd initially imagined it to be. In just my first month on the job, the provisions tent ran dry on three occasions leaving the crew on the verge of mutiny and our Sudanese cook sobbing loudly into arms hugging his knees as he squatted on the ground wailing about the shortages being the Will of Allah. And since it was haram to take issue with that edict, he intoned, there was nothing mere mortals could or should do except sit and wait for the Day of Judgement to arrive.

For him, the godless bunch of heathens he was employed to serve were keen to assure him, that day might arrive sooner than expected unless he learned that his job included doing regular stocktakes and providing us with lists of things that needed replenishing.

For all the good that warning did, they might as well have been talking to the mangy mutt that had attached itself to the crew and seemed to show up wherever we went. The Will of Allah was obviously a more powerful influence on the Curry Wallah's life than anything any of the crew could exert on him and in the end he was relieved of the stocktaking job, that onerous responsibility being loaded onto the one crew member who, because his day job involved working with figures, was considered by the other members to be best suited to the task.

If only they knew. By putting me in charge of sustenance provisioning they were in fact putting their lives in the hands of the one person on the crew with even less capacity for keeping the shelves well-stocked than the Curry Wallah.

Had my relationship with camp boss Blott not got off on the wrong foot, calls to him on the camp radio with requests to organise a re-supply might have been better received than with a string of expletives relating to the fact that I obviously still considered him to be my bellboy. Only rarely, when someone else was manning the base camp radio, did everything requested arrive and in the end it was decided to transfer the call-up job to Big Foot, the one man on the crew Blott knew had sufficient seniority to be taken seriously by head office in the US.

'Get everything we need to us pronto, Blott,' was effectively the message in Big Foot's calls to him, 'or you might find yourself having to answer to Houston re. your little sideline.'

Whether it was the prospect of a head office grilling about his HookAir activities that saw an improvement in our provisioning I know not. But I suspect it had as much to do with another little local difficulty the base camp boss found himself facing the day Blott's big mate Sitting Schmuck made a rare trip into the bush to find our beer cellar as dry as his own parched throat. That, I think it's fair to say, did not go down particularly well. Not well at all.

Chapter 2.3

If there was one thing Sitting Schmuck could not live without it was alcohol. Having been brought up in a part of Rhodesia where drinking was the national pastime and beating the natives a national tradition, being deprived of the first inevitably led to an upsurge in the second as Schmuck was to prove the day after arriving to introduce me to the delights of geophysical exploration survey work.

Not one of the Somali labourers accompanying us up the line to assist in my education was to escape Schmuck's seething, often violent, alcohol-deprived wrath and by the time we arrived back in camp their mood was so black I began to think about putting The Gravedigger on alert in case his presence was needed to box up the vile man's machete-butchered remains.

In the end it never came to that. But the following day it was clear we had a full-on down-tools strike on our hands, all of which could have been avoided if Blott had had the sense to ensure the camp was well-provisioned for the visit of the 3S surveyor-in-chief.

In a rare moment of hatchet burying between the crew's usually acrimonious Somalis and Ethiopians, a joint delegation arrived as we were breakfasting to demand Schmuck be repatriated to base camp forthwith and thence be frogmarched to the nearby town's police station to face charges of racial abuse bordering on actual bodily harm.

As much to prevent a lynching as to get the labour force back on the job, it was left to the crew's most senior member

Big Foot to step in to assure our visitors that their complaint would receive a sympathetic airing on Schmuck's arrival at base camp later in the day. It was an out and out lie, of course, Big Foot knowing full well that the Confederate States-born Blott would never do the dirty on a fellow redneck nazi. But at least it achieved Big Foot's twin objectives of only losing one day of 'production' and of getting the universally loathed chief surveyor out of the hair of everyone in the field.

As the urgently-requested plane took off with Schmuck on it, the audible release of breath from the crew could be heard in Somalia. Not least my own. After one full day in the company of the man I was feeling as murderous as the Somalis, especially when it became clear that even as a raw twenty-three year-old with but a few months survey experience I knew more about the craft than he'd learned in his forty-something years on the planet. His entire repertoire of survey knowledge had been 'taught' to me in less than an hour and for the remainder of the day I employed my new 'skills' keeping the geophysical line going in the right direction while Schmuck lounged in the shade of a spreading acacia tree thinking up new racial slurs for the survey gang's increasingly irate labour force.

By the time we had to pack up and leave to avoid having to navigate back to camp in the pitch black in a Landrover museum piece with no working lights and a misfiring engine, the labour force wasn't alone in hoping this horrid slovenly man would find himself a patch of quicksand to wander into. There are only so many lazy kaffir references one can take before one's patience snaps.

So Big Foot's intervention next morning was received with some relief all round, not least from Schmuck himself. By early evening he'd be back in the air-conditioned luxury of the base camp bar replenishing his blood/alcohol level and slapping backs with Blott as the pair reinforced one another's views on those who hadn't been fortunate enough to be born white while drinking the health of a God that'd clearly created the pair in his own image.

Standing watching Schmuck's departure in the Porter light aircraft used to supply crews in the field when Blott didn't have need of it for his hooker transfer taxi service, I wondered whether he and Blott would be able to bear including white men like me in that image category.

Only under sufferance, I concluded. Although qualifying on the skin colour front, in all other respects the relationship between us was the rough equivalent of that between pig breeder and pig. While there's a high degree of mutual interdependence between the two, you wouldn't necessarily want to eat with one.

The look on Schmuck's face when confronted with his new 'trainee' – a Brit with hippie leanings dressed in a t-shirt and shorts two sizes too big for him borrowed from chief mechanic Whackamole in the continuing absence of his own clothes (bar the underpants Whackamole had also grinningly offered) – confirmed that was exactly the thought going through the Rhodesian racial supremacist's mind. To earn his keep, he'd do the necessary and get his charge to the stage of surveying the line without supervision, but that was all. The less time he had to spend in the company of the ragtag newcomer with a liberal attitude towards his non-white co-workers the better.

The feeling was entirely mutual and by the end of my 'training' session I'd come to the conclusion that no analogy was more apt than the pig breeder/pig comparison. Besides the feeling that that's how Schmuck viewed me, you could train a pig to do the job I was being asked to do. Using equipment so antiquated it might have been used by you yourself, Sir Isaac, the only requirement was to keep the cat skinners forging ahead on a compass bearing while noting down elevation levels at fifty metre intervals.

All distances were measured using a highly inaccurate heat-affected metal measuring chain (once the Somali labourers designated to use it had untangled the mass of links from the heap it'd been left in at the end of the previous day's work) and because there were no known points to close out on at regular intervals with the quality theodolites we weren't supplied with, it was anyone's guess where the hell on the surface of the planet the geophysical line actually was. From the figures I gave them, those plotting the line's progress in whatever nice comfortable air-conditioned offices they were based in certainly wouldn't.

I'd have mentioned this to Schmuck but wasn't at all sure he'd understand... either professionally or personally. From what he'd let on about his survey experience to date something told me I could expect little more than a glazed expression in response to any mention of position-fixing triangulation or closing out and a rather more animated reaction to any inference relating to his 'trainee' knowing a sight more about surveying than he did.

Big Foot, though, was a different matter. As the crew's most senior and most cerebral member by comparison with the rest, he would surely want to know about the position-fixing deficiency. It had a potentially crew-threatening implication I felt duty-bound to draw his attention to once Schmuck's plane had disappeared from sight. Without knowing exactly where the hell we were, I put it to him, there was surely a very real risk of getting too close to the Ethiopian/Somalia border for comfort. We'd been ordered to stay a minimum of fifty kilometres away from it but without a much clearer idea of where we were and with international boundary markers few and far between in such desolate places, for all we knew we could already have crossed the thing.

That, I pointed out probably a little bit too teacherly in the circumstances, would hardly benefit the operation. Apart from attracting the attention of all the wrong people it'd put the whole survey at risk and might even spark an international incident. And the last thing I personally wanted was for that to happen. Even if we did escape a lengthy period of incarceration in a Somali jail such an aberration would most definitely result in the second worst outcome – my repatriation to the unremitting gloom and deprivation of a UK economic slump that was showing no sign of easing.

Despite the primaeval conditions here, I confessed to Big Foot, I did at least have a job and a salary coming in (even if, as Big Foot grinningly pointed out, it was a fraction of what I'd be pulling down if I'd had the brains to sign on in Houston rather than London) and I wasn't the only one facing the employment scrapheap if this operation had to close up. The same applied to the local labour, the twenty or so Ethiopians and Somalis employed as bulldozer drivers, mechanics, survey team helpers, camp organisers and general dogsbodies.

As Big Foot was about to respond, I stopped him. It'd just occurred to me that it didn't actually need an international incident to bring the entire operation to a halt. The same would apply if just his precious computer truck was impounded by the Somalis. Without it, the whole operation would have to close up. So maybe, I suggested, we might consider giving the location determination issue a bit more thought?

As my furrow-browed exposition unravelled, Big Foot first stopped grinning then started staring fixedly at the ground. Only when it had run its full course did he lift his eyes to roll them exaggeratedly in their sockets.

'God preserve us from greenhorn rookies,' he muttered to the mess tent's ridge pole before directing a dagger-filled gaze straight at me. 'Think I don't know all that, Professor bloody Einstein! You're not the only one round here with a brain you know. Maybe the only one with two,' he said, unable to stop a look of amusement at my expense flitting briefly across his eyes, 'but you only need one to work out what'd happen if we got within howitzer range of the Somalis. That, "Professor", is why we've got a plan to prevent that happening!

'Like you,' snarled Big Foot to the suddenly silenced rookie in his presence, 'there's not a man on this crew not carrying the location concern around in his head. They might not show it. Far too pussy for these knuckleheads. But it's there. Every one of them knows that from the ground there's no way of knowing how close we're getting. So, Mr Two giant fucking Brains, we take to the air to do it!'

Out of nowhere my mind's eye was suddenly filled with a vision of trench warfare being fought at the Somme or Flanders, First World War battlefields where frontline troops would risk life and limb by going up in tethered balloons to spy on enemy positions. Surely, I thought, no one would be mad enough to do anything similar out here in the middle of Africa. One mistake and it wouldn't be just the vultures spiralling around up there in the thermals.

Not quite, Big Foot assured me, but close. From time to time the Baron would order the supply plane to fly a bit further on from the end of the line's position to scout out the ground ahead and report back. No sign of troops on the ground and the crew was ordered to press on. Do a rapid ninety degree turn if there was.

'Sounds a bit precarious,' I managed on recovering some poise from the geophysicist's onslaught. 'What if there are troops and they have anti-aircraft guns? Someone might be tempted to take a pot shot at the aerial invader.'

'That's why the Baron sends his mad bastard assistant Fishball to do the job. All flyers in this neck of the woods've got a screw loose. But Fishball is one of a kind. Loves that sort of challenge… providing he's got an extra set of eyes in the co-pilot's seat.'

'Who'd be stupid enough to volunteer for that?' I said. I'd heard about Fishball and the news was not good. The man had a habit of swooping in low towards the camp, suddenly emerging over the scrub without warning to scare the bejesus out of its inhabitants just to see the looks on their faces. Got it a bit wrong on one occasion, apparently, ending up ploughing into the scrub at the end of the airstrip and shredding a passing camel to a bloody pulp in the process. He'd emerged OK but the Porter's propeller had got bent beyond use and he was forced to take it back by truck to get it straightened. Neither the Baron nor Blott was overly impressed by that, nor at having to shell out compensation to the Somali tribesman who claimed it'd been his camel Fishball had turned into streaky camel bacon.

On hearing my volunteering question Big Foot's grin returned in spades together with a mischievous glint in an eye that was now looking directly at me.

'Me? No way, Jose! You won't get me volunteering to become target practise for the Somali army.'

'Not actually a case of volunteering,' said Big Foot through a full-on orthodontal display of maniacally grinning teeth. 'Taken a look at your contract recently? You'll find it says it's the responsibility of the line surveyor. Not necessarily to get shot out of the air. Even the bunch of shitforbrains cowpokes employing us would stop short of requiring its employees to put themselves in danger to get the job done. But definitely the scouting out of the bush beyond the line bit.

'So best get used to the idea, ma man. It's all part of your job… and by my calculations we should be doing the next scout any day now. Not prone to air sickness are you?'

'Air Sicarus, perhaps,' I muttered back, forgetting for a moment I wasn't in the company of a classics scholar. In the context, the few folks I knew back home with such learning might find the allusion to the mythological figure who flew too close to the sun mildly amusing. Not here. The only likely response I could expect from the scholars at the University of the Ogaden would be a slap for being a smart arse and the look on Big Foot's now unsmiling face confirmed it.

'That does it,' the look said. 'Since you've only been here a short while I was half prepared to let you off this once and go myself. But after that…'

***

Big Foot wasn't joking. Unable to check my contract buried in the bag the airlines had excursioned into the ether somewhere between London and Addis Ababa I had no choice but to accept the punishment my Canadian 'friend' had arranged for that little aberration and climb aboard the Porter that'd flown in two days after our little talk.

Within minutes of its arrival I was bouncing reluctantly to the end of the airstrip Schmuck had had no part in cutting seated alongside the wild-eyed Texan ex-crop duster Fishball Finnegan and praying to any god that was listening that the flight would not only prove uneventful but within minutes we'd be safely back on our bit of terra firma, not crawling out of a mangled plane wreck somewhere in Somalia's.

Fat chance. Under orders from Big Foot to give me the full lie of the land scouting experience, the captain of my flight roared headlong down the airstrip without requesting that passengers fasten their seatbelts and refrain from smoking, took off almost vertically, banked sharply and dive-bombed the camp wiggling his wings as he did so while shrieking with laughter. Had he had the aircraft to do it, I'm pretty sure he'd also have thrown in a barrel roll for good measure, just to see how pale his one and only passenger would go.

As it was, with all blood drained to my boots anyway, the crazy mad pilot seemed content to refrain from any further aerial acrobatics, restricting himself to a tight circle round the camp to pick up the line before straightening out to follow it to the point the cat skinners had cut up to then staying on that bearing as we flew on into the unknown.

Strangely, with Fishball keeping an unexpectedly responsible eye on the bearing and me noting down anything on the ground that might hamper the cutting of the line, I actually found myself relaxing. Providing the Somali army wasn't better camouflaged down there than expected, it looked like the only form of life up ahead was the odd camel munching on thorn bushes. If Fishball could avoid bumping into them on landing, it looked like there were any number of decent places to build both a camp and an airstrip when the time came for the next camp move.

Fishball agreed. Or rather I think he did. The combination of a Deep South Texan drawl, his trademark chewed cheroot and the noise of an engine so inversely proportional to the Porter's hairdresser car size it'd have been sonic competition for the Starship Enterprise made it impossible to tell what he was saying. For all I knew he could have been insulting my mother or telling me to bail out. Without headphones there was no way of knowing. So I did what any well-mannered Brit would do in the circumstances and simply responded with nods and smiles and the odd thumbs up.

Had I known what I was agreeing to I might have been a tad more circumspect. Back at camp rather later than hoped it was made clear to me that the message I was apparently sending was that it was fine if we did a bit of a detour to buzz crocs in what remained of the semi-permanent Wabi Shebeli river that crossed the Ogaden.

Fishball hadn't needed a second invitation to fly miles out of our way to dive-bomb the unfortunate creatures lounging oblivious to airborne incoming on the banks of the bits of river that hadn't yet to succumbed to the dry season. No other observer he'd had in the co-pilot seat had ever given him the go-ahead to carry out this death-defying stunt and by the time we got back to camp he was giving a clear indication he regarded me as a friend for life, convinced I was as much a hedonistic existentialist nutcase as he was himself.

That's not the description he'd used, of course. The words chosen to describe me went roughly along the lines of a 'reg'lar good ol' boy bronco bustin' motherfucka' and came as a bit of a surprise to the rest of the crew who'd had me down as a reg'lar greenhorn rookie so wet behind the ears they dripped on his boots.

Now, after Fishball's comment and seeing me emerge from the

Porter with a smile on my face they seemed willing to revise that view. I'd clearly relished my flight into the jaws of hell with a man everyone knew had the death wish of the average alligator wrestler and the new looks on their faces seemed to signal renewed respect.

If only they knew. What they'd taken as a smile of thorough enjoyment of the experience was in fact one of massive relief, one that in fact my fellow crew members were also now feeling. If I was OK about accompanying Fishball on his terror flights into the unknown that meant they'd never be called upon to fill in.

That finding was almost as comforting to them as the news I returned with. With no sign of either the Somali army or any great geographical obstacle up ahead they could now relax for a while. 'Normal' service could be resumed without fear of imminently falling foul of any unexpected minefields and for the moment we could all rest easy in our beds… providing the lions didn't start showing an inordinate amount of interest in us. The one's the Somalis said they'd seen prowling close to the camp.

Chapter 2.4

Nice try Somalis. You might have had me worried but not the crew's more experienced members. This, or something very much like it they told me, was a regular ploy tried on newcomers by the labour force trying to get a day off work.

'Fear not,' those who didn't have to spend hours alone in the bush peering through the lens of an antique theodolite with their attention directed away from what was going on behind them assured me, 'it's just the Somalis crying wolf. Or in this case lion.'

How very reassuring. Even so, it still seemed prudent to have one of the team covering my back while up the line and there was even a remote possibility of some large pointy-toothed predator creeping soundlessly up to take chunks out of me. At least that'd mean it got him first, not me.

To my relief, in all my time working as an exploration surveyor in the Ogaden the urgent tap on the shoulder never did come... except once.

'Mr Mark,' my trusted lookout Abde murmured in my ear a couple of weeks after the 'sighting' of those lions. 'Company.'

Picture a mammoth locked in ice, a beetle immortalised in amber or the residents of Pompeii in the wake of that Vesuvius eruption and you'll have some idea of my degree of mobility on receipt of that news.

For the first time in my life I understood the true meaning in the term 'petrified'. Every muscle froze, every sinew tensed and every sense soared, trained exclusively on detecting any movement or noise or unusual smell in my immediate vicinity.

Then I heard it. The dreaded low rumble of a large beast breathing almost down my quaking neck.

What to do? Remain stock still or take off like a jack rabbit dodging and weaving and not looking back to see how close I was to becoming the main course on the visitor's menu.

In the end it was my sense of hearing that decided matters. Unless those TV animal programmes had missed out a lion's ability to mimic other beasts this was no visiting king of the jungle. More complaining grumble than deep throated growl, the announcement of its presence could only belong to one animal – a fully grown Ogaden desert bloody camel.

Turning cautiously to check my hearing hadn't deceived me I was met with a grin on Abde's face that'd have lit up a moonless desert night.

Staring me full in the face and just inches from it was said camel and its equally amused owner holding out a large wooden bowl filled to the top with milk still steaming from his humpback companion's insides.

Once more, the what to do question arose. Should I gratefully accept or politely decline the invitation to sup from the bowl? Having only been in Africa a matter of weeks, camel milk had not knowingly passed my lips and apart from having no idea whether I'd find it palatable there was the small matter of the health and hygiene aspect to consider. For all I knew the bacteria in unpasteurised camel milk could be responsible for the deaths of millions.

In the end it was Abde who broke the deadlock. Noting my hesitation, his eyes pleaded for me to take it. Refusing would be taken as a serious insult, they said, and considering who was offering it might result in something equally as health-threatening as stumbling on a hungry lion. 'This is no ordinary Somali camel herder,' the eyes said. 'See what he's carrying?'

Ah. See what you mean, my eyes messaged back. Apart from

the usual long sharp slashing knife tucked into the belt of every Somali nomad, slung over his shoulder was a World War Two Lee Enfield 303 rifle which I doubted he'd be averse to using. In fact, I knew he wouldn't. The hole in the wing of our Porter light aircraft was evidence of it.

On coming to a thankfully safe halt on the airstrip of the third camp I'd had a part in selecting and building, Fishball sprang out of his seat and was racing round the plane to inspect the damage before I'd unbuckled my seat belt.

'Motherfuckin' asshole camelturder!' he was cursing by the time I caught up with him. 'See what that shitbag motherfucker did?'

At first I could see nothing, the hole Fishball was pointing at obscured by an accumulation of filth that seemed too much of a bother for anyone to ever clean off.

I did when he put his finger in it. A neat round punch mark right through the wing that could only have come from the sort of rifle my camel herding visitor was carrying.

'Sonofabitch! Now THAT'S what I call shootin' dude! None of your army compadres would be able to do that. Hit a plane five hundred feet up flying at ninety miles an hour! Fuckin' A!'

Despite my best efforts I had trouble matching Fishball's admiration of the marksmanship. While it was undoubtedly an impressive piece of shooting I wasn't sure I'd be especially in awe had the bullet impressed itself in me in just my third reconnaissance mission over no man's land.

Up to that point I'd become almost lackadaisical about going up with the crazy Texan. Once he'd got showing off his flying acumen out of his system in our first terrifying flight together the second was altogether more uneventful and by the time I was due to go up on the third I was almost looking forward to it.

Taken in the right attitude, I told myself, these missions knocked spots off the regular daily grind. Once you'd done one day of line surveying with equipment only marginally more advanced than a compass and a long bit of string, arguing the toss with the cat skinners over the direction they should be heading, abandoning all survey work when the heat haze got too bad and queuing to de-cake yourself of bull dust in the camp's single dribbling shower before collapsing into bed with a bellyful of beer, you'd pretty much done them all and any departure from all that was a welcome change.

So when the time came for a repeat flight over virgin scrub I found myself counting the days. Now dressed in my own hot weather clothes after the miraculous recovery of my bag from its excursion via Nairobi and Dar es Salaam, I was looking forward to this time, I hoped, spotting some of Africa's famed wildlife herds. I'd been in the continent a full two months and all I'd seen so far was an enormous and clearly very old giant tortoise, a camel rather too close for comfort, a scorpion in the shower – ditto – those crocs in the Webi Shebele – double ditto – and a spider the size of a dinner plate high up in the mess tent positioned strategically above Sitting Schmuck's head and about which we thought not to warn him for fear of giving cause for concern.

Where the vast herds of zebra and wildebeest TV wildlife programmes showed sweeping majestically across the plains? Where the families of elephant and giraffe grazing contentedly on the branches of Africa's tallest trees? Where the prides of prowling lions the Somalis kept reporting but which seemed shy of making themselves known to anyone else? Not where I was looking, that was for sure. Even from the air.

With a clear view of the landscape from up there, anything that moved would've been spotted. But in all the flights I took not once was there anything to report back. The scorched landscape was deserted save for the odd emaciated camel wandering forlornly amongst the thorn bushes and, said Fishball, therein lay the answer to the question. An unremitting drought like no other he'd seen in Africa had left all wildlife either running for its life or shrivelling to death under the unrelenting sun. And the same applied to the Ogaden's nomadic tribes peoples.

I was to see what he meant the day I had to return to base camp to sign the receipt for my recovered bag. With the plane doing a HookAir run that day and our squad of Ethiopian army 'protectors' going in to be replaced by a fresh squad, I hitched a ride on the back of their truck and found myself part of one of the most unexpected humanitarian rescue exercises I could have imagined.

About an hour out from Kebri Dehar we passed a skeletal Somali camel herder squatting exhausted by the side of the track, no camels in sight. Considering the historic antipathy between all Ethiopians and Somalis, to my amazement the army truck stopped and the platoon commander asked the man if he needed any help.

With a particular antipathy between local Somalis and any Ethiopian deemed to be part of the force occupying what the locals considered Somali soil, the herder thought hard before answering. This could be a trap, you could see him thinking. One designed to win his confidence and perhaps con him into providing information on Somali military positions in the region.

On the other hand, considering he was on the point of death through dehydration, what choice did he have? So, almost reluctantly, he summoned up what strength he had left and slowly held out his dessicated goat skin water flask. Not a drop remained and it was clear it'd been that way for some time. The flask was as cracked as the mud of a long dried up water hole.

Once revived with water offered by the platoon commander, the camel herder's tragic story gradually unravelled. Every one of his small herd had either died or wandered off into the bush and he'd spent the best part of a week trying without success to track them down, continually aware that unless he found them soon his own fate was sealed. With no water left in any of his territory's water holes, until it rained he and everyone like him were completely dependent on milk from their herds.

Listening to the commander's translation of the herder's story, I felt a wave of guilt-tinged enlightenment flooding through me. Without having any need to, I'd supped from the milk bowl of my rifle-toting herder friend just days earlier. Could the one mouthful I'd swallowed with a grimace and a pledge to avoid any repeat experience be the one that stood between life and death for HIM? By just being there, had I deprived him of something on which his life depended?

The answer, I decided, lay in that look in Abde's eyes. Thinking back on it, I realised they'd been telling me a lot more than I thought at the time. They told me the man had no option but to offer me the milk and I had no option but to accept it. In Somali tradition, apart from it being as taboo to refuse an offer of hospitality as it was to not offer it in the first place there was the small matter of gift reciprocation. A gift taken had to be matched with one given and, Abde's eyes had said, this was the only gift the herder had to match the one I'd been instrumental in giving him and his family.

It took a while, but in the end I worked it out. I – or more precisely the crew – HAD given them a gift. Several, in fact.

Prior to every camp move, so as to be able to load the crew's gigantic rubberised water storage bag onto a truck, the bag first had to be emptied of what remained of water drawn from wells Digit and his team drilled near every camp. At such times, Somali women appeared out of nowhere to fill any receptacle they could lay their hands on with what to them must have seemed a gift from the gods and carry the precious, life-preserving liquid off on their heads.

Being usually up the line doing my day job at the time the bag was emptied only once did I witness this joyous cacophonous spectacle. But it was a moment I'll never forget. For that was the moment I set eyes on absolutely the most beautiful knee-trembling woman I've ever seen, before or since.

Milling around with all the other gleefully ululating women was a vision that left me speechless. Arrow-straight and taller than the rest, even at fifty paces I could see how her sky-high cheekbones, perfectly aquiline nose, coal-black eyes, luminous white teeth and flawless dusky complexion would have men anywhere worshipping at her heavily bangled ankles and I rushed to get my camera before she dematerialised back to whatever spirit world had sent her.

To my relief she was still in earthly form on my return and I began creeping closer to get a better shot.

It was not to be. For no matter how surreptitious my forward movement the distance between us never varied. Without apparently moving her feet, the closer I got the further away she seemed to float and I began to wonder if I was being treated to some supernatural force in action. A force only those indigenous to Africa knew how to harness.

Without once looking in my direction she maintained a respectable fifty pace distance and it soon became clear that without a powerful telephoto lens she'd be little more than a speck on the landscape of any photograph taken.

I could imagine her smiling on seeing me eventually giving up and wandering away shaking my head. 'You'll have to do a lot better than that to immortalise me on film,' I could hear her thinking, 'starting with not being such an obvious novice when it comes to understanding Africa and the Africans. Having been here no more than a microsecond, fancy thinking you could creep up on one without them noticing! I'm embarrassed for you.'

She was right and it was a lesson well-learned. Apart from the obvious fact that you had to be in Africa a lot longer than I'd been to get anywhere close to understanding what was clearly a land of far greater mysticism than I'd been led to believe, in that moment I realised for the first time that education wasn't confined to institutions. The wide wide world could be every bit as educational as any university providing the student entered it with an open mind and was prepared to change the habit of a lifetime and start seeing as well as looking.

Until now, I'd had enough to do just adjusting to my new life. But now I was bedded in, the incident with the Somali angel and the water bag convinced me the time had come to shelve all preconceptions and start the seeing process in earnest.

Looking back, it's a wonder I didn't choose to go blind instead. For once tuned properly in to the desert environment I discovered there was a lot more going on here than I'd imagined, almost all of it life-threatening. From sand flies to spiders, scorpions to snakes anything that could make your life a misery was here and woe betide anyone who treated these local inhabitants lightly. Anyone who did would instantly become educated to the literal meaning in the term 'come back to bite you', a phrase I'd have been able to add to my own growing catalogue of life experience had some sixth sense not kicked in to save me one day out on the line.

Taking a break from the eye-straining squint through the ancient theodolite's clouded lens, I was about to settle myself in the welcome shade of a nearby acacia tree to rehydrate when something stopped me. 'Not a good idea,' some voice was intoning in my head. 'Take a closer look.'

Blinking hard to clear my vision from all that lens peering, I saw what the voice meant. Right where I was about to park my arse was a seething mass of fire ants just goading me to try my luck. One step closer and I'd have been covered in the voracious things stinging and biting and letting me know in no uncertain terms that man was not, as he thought, the master of this particular patch of the universe.

A recounting of this close escape to my fellow crew members as we hacked our way through the usual evening meal of incinerated steak and cold oily chips did not elicit the response expected. No gawps or gasps. No oh's or ah's. Just a collective blank stare and the odd pitying shake of the head. Not at all how I'd thought they'd react. Was it something I'd said?

The answer only came as I made my way back to the mess tent from a pit stop made necessary by the third beer downed in the space of twenty minutes. As I approached the tent, I picked up the basics of an exchange between Whackamole and Big Foot which explained everything.

'… started seeing things,' Whackamole was opining. 'Seen it before. Leave 'em out here too long and they start thinking they're growing antennae. All that crap about sixth senses. Sick sense more like. The man's showing signs of going bush. Maybe time for a word, Footsie… before it spreads.'

'Yeah. Could be right,' Big Foot muttered back. 'Maybe time for an oil change. I know I need one now and again and he has just done a treble shift. I know that'd get to me. I'll have a word with Blott but I ain't looking forward to it. You really want Schmuck back here for a whole two weeks while he's on leave?'

Chapter 2.5

I think it was all the stuff about the desert being as alive as Piccadilly Circus that did it. To those who'd had some time off to de-bush themselves, such musings did sound like the ravings of one who'd been left out in the sun too long. That's certainly the way I viewed my fellow crewmates on my return from each period of leave.

So maybe Whackamole had had a point. Maybe nine weeks straight in the field had sent me a bit bush. The only way of telling was to get away and see how Addis Ababa had been faring without me since my last spot of oil change time off. Blott had grudgingly agreed to Big Foot's recommendation to let me go despite all forebodings over having to put up with Schmuck until my return and the base camp boss had given me an option. Take the full two weeks leave owing or keep it to one and win yourself double pay for the week's leave you give up.

Cheeky bastard. Fancy thinking he could get that one past me. While I was no mathematical Einstein even I could work out that on the basis of three on, one off, after doing nine weeks straight I had three weeks leave owing, not two.

Was that actually a look of grudging respect on Blott's face when I pointed this out to him? Maybe. But more likely one of smoking suppressed rage over having been caught out by a complete novice to the oil-hunting game.

He did his best to hide it, of course, but Blott was nothing if not transparent. That look on his fat-jowled beetroot face also told me that from now on I'd do well to watch my step. Every move I made from here on in would be under his microscope being inspected for any small aberration, any minor mistake with which to beat me and square the series… especially after the deal I managed to prise out of him.

'OK,' I said. 'I'll take just the one week's leave. But for the next three work shifts I want them to be on a two week on, one off basis. Not three weeks on. That'll recoup the time you owe me.'

'But I'd only owe you two weeks leave if you took just the one week now,' he scowled. 'Where d'ya get the three shorter on periods from?'

'From the time off earned while I'm doing the first two two-weeks on shifts.'

'You limey motherfucker,' I could see him thinking. 'Getting me once earned you a top-ranking place on the watch list. Twice? That REALLY gets you noticed. You sure you wanna risk that?'

Frankly, my face told him back, now I've seen the scale of shoddiness of this operation I'm not sure I give a damn. So take it or leave it shitbag. Your loss if you don't.

To my amazement he did and by the end of the day I was on the DC3 heading for Addis with a fistful of per diem allowance, the thirst of a dying dingo and the hope that the first would be enough to cater for the second. The last thing I needed was to have to dig into my salary to supplement my stay. I'd need it all if, as Blott's eyes suggested, my little 'win' over him ended in my being 'suitcased' and sent packing back to a UK that showed no sign of escaping the spiral down into the full-on economic slump the soaring price of the very commodity I'd come in search of had sparked.

***

Once one knew the ropes, Addis Ababa was a hairy-arsed oilman's dream. As my first period of leave following my first three-week stint in the field had taught me, there seemed nothing that couldn't be got – up to and including a willing 'companion' with every beer, any automatic weapon you thought might augment your wardrobe and the rarest of endangered animal hides if that was your thing.

In my case it wasn't, much to the irritation of the illicit goods sales gangs frequenting our favoured bars and who one learned to steer clear of lest it became necessary to 'encourage' them to try their luck elsewhere.

That, of course, only came through experience, the learning curve going almost vertical the first night of my first leave when the little gaggle of bush escapees I'd joined up with were befriended by a couple of local barflies cajoling us into experiencing a 'real' Addis hang-out.

After checking into the dive Hotel Afrique most of the crew stayed at to make their per diem last the distance, first stop was a bar close by whose idea of opening hours was dictated by the presence of anyone whose thirst Ethiopia's not unpalatable local beer had yet to quench.

As one a.m. approached and with the bar starting to thin out, our newfound friends suggested we might like to extend the evening with a trip to a club specialising in Ethiopian 'culture'. Nothing if not eager to expand our cultural education into the area of local traditional arts and crafts, we enthusiastically agreed and after a short taxi ride around the Ethiopian capital's crumbling unlit backstreets found ourselves being led up a flight of stairs behind a shop stuffed to the gills with leopard skin rugs, bootleg whisky and boxloads of cassettes of pirated well-known albums.

Driven on by some anthropological urge to discover what cultural delight lay in wait for us at the top of the stairs we almost bounded up them to find ourselves seated round a rickety table in one of the club's most light-deprived corners being plied with the sort of cocktail you placed a lit cigarette next to at your peril.

As the cocktails went down and the noise level of the club's drum and unidentifiable string instrument band went up, a single hard-beamed spotlight flared to pick out a group of dancers stamping and ululating in time to the music. Warming to their task, the dancers' routine first morphed into full-on anarchic fling to all corners of the dance floor before gravitating back to the centre to form a heaving, semi-naked mass of animal skin and feather clad humanity.

There the stamping and yelling resumed for several minutes before the mass suddenly parted to reveal a heavily-armed tribal warrior gesticulating ferociously at us in full battle cry mode – a sight so terrifying it prompted a scene from the film 'Zulu' to make an unexpected appearance in my drink-addled mind.

In that moment I suddenly knew how the company of red-coated British troops depicted fighting at the Battle of Rorke's Drift in the film must have felt on being confronted by thousands of such warriors charging at them. Knowing what was coming they'd have been bricking it and praying that the end would be both swift and painless… just like I was doing now.

I wasn't alone in the thought. Glancing around I could sense every one of our booze-fuelled party tensing for an assault by this warrior with the whirling blades, drinks frozen halfway to their mouths, their short-lived lives flashing through their hooch-misted minds oblivious to the grins on the faces of our hugely amused Ethiopian hosts.

Mission accomplished, the grins said. Looks to us like you're now starting to understand where you are. This ain't any old African country, lads. This is Ethiopia. A place where, unlike those parts of the continent that hardly resisted being colonised, we fight back and have been fighting back for centuries… as Mussolini found out to his cost when he tried to occupy us in 1935. So if you've got any thoughts of doing anything similar our advice would be simple. Forget it. Unless, of course, you want to find yourselves being repatriated in several small boxes. Clear?

Damn right, our sweating ashen faces said in return. Not

only is the message received loud and clear but be assured. As ghostlike as we must look to you we're not Mussolini spectres. We're just common working men here to do a job, and if it looks to you like we're a bunch of neo-colonialists on the make, here to plunder Ethiopia's resources by the back door, we apologise. That's certainly not our intention. Hopefully, your country will get as much out of our work as we do.

Would that satisfy our hosts, the paranoid, suddenly more sober glances we threw at each other said? Or would they see right through what we all knew to be a blatant untruth and set the warrior on us to administer the last rites?

Either way, the glances said, I think it best if we got the hell out of here. Those blades are looking more razor-like by the second.

'Don't worry,' our hosts had begun to assure us as all this message-sending glancing was underway. 'He means you no harm. It's nothing personal. Just a dance the Afar do to celebrate a successful raid on the Oromo.'

Shee-it, I thought. Didn't the Afar get a mention in Danny Boy's interminable horror story blather about Ethiopia and its level of tribal savagery on the way out here? If my addled brain recalled correctly, didn't he say they were the most brutal warlike tribe in the region? The ones to whom the word 'prisoner' was an alien concept?

If there was even a modicum of truth in that, regardless of our hosts' smiles and reassurances about our safety it was most definitely time to make ourselves scarce. Prisoners would be the last thing we'd be if, on thinking about it, they came to the conclusion that resource-plundering neo-colonialists was not only exactly what we were but the worst sort of resource-plundering neo-colonialists. The type that regarded those being colonised as so backward and stupid they wouldn't twig they were being lied to.

'... or at least it won't get personal if you'd be good enough to consider a donation to the dance troupe through a purchase or two at the shop downstairs,' our hosts' eyes were adding. 'No offence, but without such a donation I'm afraid I can't provide an assurance that the man with the blades will be able to fully resist demonstrating his prowess with them.'

Gottit, the eyes of every one of our little band of quaking brothers flashed at one another together with one final, urgent message – run like fuck and don't stop running 'til you're safely away in a taxi!

***

Waking up in the company of the mother of all hangovers the next day, the full horror of what we might just have escaped kept coursing through my pounding head. Would it REALLY have come to being diced and sliced and served up on an Afar barbecue?

Surely not. All that stuff about unreconstructed savagery in Africa was surely just blarney put about by people like Danny Boy to scare the uninitiated. Although there wasn't much rule of law here there was some and your regular African would surely have sense enough to see they'd never get away with the kind of treatment their ancestors would've doled out to anyone crossing them.

On the other hand, having been here the full sum of three and a bit tempestuous weeks what the fuck did I know? Maybe such acts of dicing and slicing weren't just commonplace in such under-developed parts of the world, but the norm. So perhaps, just to be on the safe side, maybe best to stick to slightly more refined establishments for the time being? Just until I got to know the place a bit better.

Anyway, even as a reasonably fit twenty-three year-old I needed time to recover. Sprinting away from the scene of the crime along blacked out alleyways with no idea where we were headed and a blood alcohol level to test the limits of the average breathalyser would have left me debilitated at sea level. At seven thousand feet up in Addis, even without the added handicap of a bellyful of hooch I'd have had difficulty complying with any order to blow into said breathalyser. With it, the chances of being able to oblige were in the vicinity of zero.

So taking refuge in somewhat more genteel surroundings for the foreseeable was decided on and as night fell I was boarding a taxi en route to what I'd heard other members of the oil worker retinue refer to as their sanctuary from Africa.

Stepping through the door of the quite civilised by comparison Cottage Pub later that day it was clear I hadn't been the only one deciding they needed a moment to get their breath back. Almost the entire complement of last night's outing was there trying not to mention our helter skelter flight from hell. No one wanted to have a pussy label attached to them by those with long enough African track records to have already amassed a healthy collection of been there done that t-shirts. Dawg especially.

In town with a cohort of other Stetsonised shitkickers from the company's other crews it didn't take long to discover why Dawg and his ilk had dubbed The Cottage their base camp. No queuing at the bar for them. Just a nod in the direction of the Ethiopian barman as they helped themselves to the booze stock and carried their bounty off to the 'top' table – the one occupied by ranks of huge-haired blousy women with equally expansive blousy bosoms, exactly similar tans and the type of smile you only see on adverts for denture fixing products.

We had, it seemed, arrived on family night. The one on which the company wives were allowed by their Texan driller husbands to accompany them out on the tiles providing they didn't expect to be treated as anything more than one of the good ol' boys.

By and large it was clear they were happy to oblige. From what I picked up from their chatter, these outings were a welcome break from the oilman wife's regular routine of trips to the pool and the beauty salon, 'coffee' mornings with the other wives and ordering local skivvies around at their sprawling palatial ranch-style homesteads dotted around the Addis Ababa outskirts.

Only when their men were in from the field would the routine vary and was something every one of them obviously looked forward to with relish. None would ever so much as think about hitting the town without their male chaperones. In Texan oilman society such behaviour was clearly viewed as being about as acceptable as being caught ordering salad to go with the never-varying oilman diet of elephantine t-bone steaks and ketchup-swamped French fries.

While those of our gang who were familiar with the Texas brigade joined in with the whoopin' hollerin' good ol' boy shitkicking revelry, the Cottage first-timer entourage took refuge in a dark corner to nurse the beers we seemed to be the only people paying for.

But not for long. Cajoled by their wives to bring the 'cute new rookies' into the throng, Dawg and his lookalike band of caballeros were soon at our table flexing their deep-tanned muscles and challenging us to prove our manhood with what they termed Texan-style party games.

Unwilling to provoke either Dawg's wrath or the company wives' derision, we had no option but to oblige and soon found ourselves inveigled into joining in with a series of 'games' that turned almost exclusively out to be trials of raw unadulterated brawn. From basic arm wrestles to fingertip push-ups to solar plexus punch resistance we jointly earned zero admiration from our drill pipe bending opponents, about as much sympathy from their shrieking brides and were about to run up the white flag when someone suggested a game of darts to round the games off.

'C'mon y'all!' drawled Dawg. 'Not chickin' are ya? Ev'nin's jus gettin' started 'n ah'd hate to see y'all get whupped into having to pay the bar bill!'

Eh?

'Cottage rules!' beamed Dawg. 'No win, no escape. Lose ev'ry game and you pick up the ev'nin's tab! Didn't I mention that at the start? Well... dang!'

Oh shit. After the previous night's expense there was no way our gang was going to be able to cover that. The top table's cocktail bill alone would bankrupt us. So now we had no choice. It was win the darts match or go home shirtless.

Fortunately, with more non-Americans on our team than our opponents, we stood a better-than-evens chance. Unlike the UK and other non-American countries, darts had never really caught on in the US as Dawg was about to demonstrate on explaining the rules of darts as he understood them.

'OK, listen up y'all. Twenty-to-one is what we's playin'. Start with the twenty and work down to one. Y'all got five lives, one lost each time you don't hit the number you need with three darts. Lose all five lives and you're out. Gottit? We play til there's just one man standin' and it's winner take all. Let's go!'

Compared with what I and others on our team were used to back home this was simplicity itself. No need to hit doubles or trebles or the bull as in 301 or 501 down since it was clear Dawg & Co had no idea what those bars on a dartboard were for. So this should be a cinch I thought and I caught the team casting furtive glances of delight at one another. Without knowing it, Dawg had surely just thrown us a lifeline.

Yes and no. Versed in the art of pitching horseshoes in their backyards, our opponents had a mean eye despite the booze and their general technique of hurling darts like a baseball. So as lives were lost and team numbers shrank it was still evens the field as the last two standing stepped up to the plate – me and Dawg.

Down to three numbers each and both with two lives left it was going to be a tense finish. He who kept his head better was going to take the prize and thanks to the differing amounts of hooch consumed that put the odds on me. Still feeling the effect of the previous night's cocktails, I'd been a virtual teetotaller compared to Dawg who by the time we made it to the final was clearly having trouble focussing.

Believing more drink would help, he first drained his glass, then topped it up, took a hefty swig and steadied himself to throw. Result – darts everywhere but the place intended and two minutes later I was crowned 'champeen', beating the reigning champ by a single life.

Relief in our camp was as audible as the gasps of disbelief from our opponents. No one beat Dawg if they wanted to avoid the lash of being seen to be a threat to his Top Dawg status.

The eyes on me said it all. 'Yo a dead man, Two Brains. No one embarrasses Dawg and lives to tell the tale. Last guy that did found himself not only suitcased but narrowly escaping being shipped out IN the suitcase!'

Oops. Hadn't twigged there was top dog face at stake here. Just bankruptcy avoidance. Ah well. Done now. All I could do was hope the incident hadn't left Dawg's nose so out of joint that my own was now at risk. Only time would tell and fortunately Dawg's imminent return to the bush gave me a grace period to let him cool down.

Meantime, I thought, maybe best to give The Cottage a wide berth. Judging by the look on Dawg's face as he had to dip into his own wallet to cover his and his wife's bar bills there was every chance he'd be going back to work having left word that the remainder of the limey rookie motherfucker's stay in town should be made a time in his life that'd remain lodged in his memory for a very long time.

Chapter 2.6

Stepping off the DC3 onto Kebri Dehar's airstrip a few days later it took just minutes to discover that news of my aberration wasn't confined to Addis. Dawg had ensured that word to make my life a misery had also been pinned to Blott's heart and in the light of my little run-in with the base camp big cheese on my initial arrival he was more than happy to oblige. Once again, my trip out to the field was ordered to be done overland, the Porter being 'unfortunately' unavailable that day.

The only difference between this torture and my first expedition on the back of the bushwacker express supply truck was that this time I knew what to expect and was ready for it. Swathed in a tarpaulin and tucked in behind the cab atop the rear bay-mounted reserve fuel tank I reckoned I'd be able to escape being asphyxiated by the dust.

I was right but unfortunately had still overlooked one vital aspect. The reserve tank wasn't as securely fixed as it might have been and when the truck hit a rock two hours into the four hour trip it shifted without warning to trap my foot between it and the side of the truck leaving me hammering frantically on the cab roof. One more tank movement and they'd be changing my crew nickname to Peg Leg.

Holding my breath and waiting for the inevitable agony if the tank moved one centimetre more during the truck driver's equally inevitable standing on the brakes, for the third time in as many weeks after those two heart-stopping flying incidents with Fishball I found myself breaking habit of a lifetime and praying. With no one up the line with any medical training and only The Gravedigger back at base camp to treat me it was, I decided, God or bust.

He, she or it must not only have been listening but to have decided to have a little joke at my expense. For as the truck slewed to a halt the fuel tank slid back to its original position, my foot came free and I found myself groping for the means of convincing the clearly unconvinced Ethiopian truck driver and his mate that anything untoward had actually transpired. All now looked exactly the same as when we'd set off from base camp and with no one else on the back of the truck to corroborate my claim, the pair simply turned, clambered back in the cab and roared off down the track shaking their heads.

Would the incident be worth mentioning to anyone when we reached camp? Probably not. One more job loaded on them to fix the fuel tank firmly in place would hardly help when it came to winning friends and influencing people – something I knew would need prioritising after having had the hex put on me by Dawg and my expectation of being given the dirty rat treatment after condemning the crew to a week of having to suffer Sitting Schmuck's presence while I was on leave.

***

Surprise was not the word for their reaction to my arrival. Rather than making their displeasure more than plain, not only did they rush to help me down from the truck but those in attendance then proceeded to usher me ceremoniously into the mess tent where whatever refreshment took my fancy was heaped in front of me.

Fussing around me like mother hens this wholly unexpected performance had me squinting suspiciously back. Surely there had to be some sort of sting in the tail to all this. Until now, all my presence on the crew had inspired had been that rib-tickling for bringing the clap TO Africa and once that joke had worn off I might as well have been a mirage for all the recognition I'd received.

But that was before Schmuck had taken my place. Only then, Big Foot whispered in my ear, had the full value of my presence been recognised. Sobriety, decorum, diligence and good manners personified in comparison to Schmuck – who'd apparently outdone even himself for permanently sozzled obnoxious bigotry for the entire week – my absence had not only been noted but as good as mourned over.

After a moment to recover from the shock of all this unexpected brotherly love, another was quickly to follow. Even the bull dust in my eyes couldn't prevent a small tear welling up as all this newfound sincerity got to me along with a thought I never imagined ever having out here. The existence of Schmuck, I'd just realised, was actually a lot more valuable to me than I'd thought.

The polar opposite, I was ultimately to discover, was simultaneously going through the mind of one crew member whose smile seemed broader than the rest. Not previously known to me having not been in residence when I left, the man was now looking at me as if I was one of the chosen ones.

My return look was one of intrigued curiosity. This James Robertson Justice gentleman actor lookalike wasn't just a new kid on the block, he was so new he'd yet to develop the regulation crew look of shipwrecked unfortunates left to survive on a desert island on a diet of roadkill and industrial waste.

'Name's Tom,' he introduced himself in the sort of cultured accent you only developed at the most private of English public schools. 'Pleased to make your acquaintance,' he said reaching out a surprisingly well-manicured paw to be shaken.

'Major Tom, actually,' interjected Big Foot. 'You'll see why when you're out in the field with him. Only man I know who spends more time with the stars than his own family.'

Crikey. Were we in the company of movie greatness? He

certainly had the look of one to whom the film set was as familiar as his own bed.

How deceptive appearances can be, I was to discover two days – or more precisely nights – later after being ordered by Major Tom to have a good lie-in next day.

Gratefully accepting the chance to get over the truck ride here, it wasn't long before I was comatose in my tent, falling asleep without even asking why I was being given this preferential treatment. All would presumably be being revealed in Tom's good time so there seemed no point in losing precious sack time worrying about it.

By the time I woke the camp was deserted, every member of the crew gone about their daily chores so after a leisurely brunch of fried egg sandwiches and coffee I returned to my tent to continue the recovery process.

As darkness descended and the crew began drifting back in for the evening meal, Tom arrived at my tent beckoning me to follow him to one of the better-maintained vehicles, the rear already loaded with box upon box of obviously well-used but lovingly looked-after pieces of kit.

With Tom at the wheel and remaining smilingly tight-lipped as to the purpose of our mission the entire way, we bumped along the line in pitch blackness to reach its end point half an hour later.

'Right,' he said, the first word he'd uttered the whole trip in spite being bombarded with questions, 'this is where your education begins... starting with helping me unload my little arsenal.'

Holy crap. Had the man brought me on a night time hunting expedition without telling me. Was this the fabled Great White Hunter of H Rider Haggard books fame? Thinking about it he did have the look of one used to peering through a telescopic gunsight.

'Ha! No young man,' he laughed on being asked direct. 'Only thing we'll be shooting tonight is the shit.'

How very true, I was to discover after we'd finished unloading and setting up a bivouac camp complete with folding chairs, oil lamps and the wherewithal to boil a kettle over an open fire. As the water heated to give us our first brew of the night Tom was to prove as good as his word, the shooting of the shit starting with the man finally explaining his presence on the crew.

'See that box?' he said, pointing at a fine well-made and obviously much loved and much travelled rosewood chest. 'Something in there I think you might appreciate. Take a look.'

I gasped when I saw it. There, lying on a protective bed of red velvet, was a veritable prince amongst theodolites. One of the most precision-crafted pieces of Zeiss-built survey instruments I'd ever seen that made the one I'd been using look like a child's toy. It was indeed a beautiful piece of work, but what on Earth were we going to do with it at night, I wondered? All the survey work I'd ever done needed to be carried out in full daylight.

'That, young man, is going to tell us where the fuck we are,' grinned Major Tom. 'The little beauty you're looking at goes by the name of Esmerelda, my very own lady of the night with whom I've been having a close relationship for, well, let's just say more years than you've had hot triangulation days.'

Unpacked and raised onto a tripod that looked built to bear the weight of a Howitzer, Esmerelda soon revealed her extra special qualities. Pointed in the right direction and caressed gently in the right way she could pick out any specific star in the firmament and Tom had a special job for her tonight. With my help, the pair would be pinpointing to within a few metres just how close the crew was to attracting the unwanted attention of Somalia's border protection force.

***

Tom, it turned out over a steaming mug of fine Ethiopian coffee as we sat around the campfire waiting for the stars to align themselves to Tom's satisfaction, was that rarest of survey species, an astro-surveyor. A specialist in his field brought in for short periods by geophysical survey companies to fix their crews' positions on the ground, something that could be achieved through recording the positions of certain stars at a given time and subjecting the readings to a complex calculation involving various mathematical tables and, Tom said, a little bit of magic.

Listening intently to Tom's explanation of the technique, I found myself breathing out. Not only, to my great relief, would I finally know where we were in relation to the border but for the first time on this trip I was in the presence of someone with whom I had something in common. I might as well have been talking particle physics to fellow crew members when, on the rarest of occasions, any had shown any interest in learning more of the job I'd been hired to do.

So, as the night passed with Tom and I spending lengthy periods between star sightings swapping life history stories, it soon became apparent that here was a man whose career I could do a lot worse than emulate.

Here was someone who seemed to have it all. A rich history of learning his trade with the Iraq Petroleum Company until it was nationalised in 1972, subsequently hiring his specialist services out to anyone with the money to cover his not insubstantial consultancy fee. That way, he said, he could still travel the world at someone else's expense, not be marooned in remote places for longer than relatively short periods and be in the luxurious position of being able to pick and choose the jobs he took on. Organised right, he said, that meant he could not only escape the worst of every British winter but never miss one ball of the English cricket season, his one passion beyond all things astro.

Bliss, I thought. That was definitely the way to arrange things. But weren't there any drawbacks to such a nomadic way of life?

'Well, maybe one. Getting saddled with nicknames like Major Tom… as in the primary character in Bowie's 'Space Oddity', he expanded on noting my quizzical expression. 'Big Foot's little joke that one for someone who spends more time with the stars than with his nearest and dearest.'

Ah so, I thought. I'd been wondering about that little Footsie-ism. Hadn't made much sense at the time to a man whose brains were still rattling around courtesy of half a day of bouncing from side to side on the back of a supply truck.

But now I'd had it explained to me, it did raise another issue. Being a night owl who spent long periods away from home, didn't that have an effect on his family life? 'No disrespect, Tom, but once I've reached your time of life I think I'd be wanting to spend more time off work than on.'

'Well yes, it's true,' he said. 'I don't get to see as much of my lovely wife as I'd like at times. But she doesn't complain. Having me around the whole time drives her scatty. She's said so to my face and has even been known to offer my services to interested parties without asking when she thinks I've been home long enough. Go back to your mistress Esmerelda, is her way of telling me I'm cramping her style at home. So largely I comply without arguing. Best way of keeping a marriage solid I reckon. Well, that and having a real mistress, of course.'

***

As the night wore on with no further mention of mistresses, Tom switching between cooing over the night time star visibility and griping long and loud about having to repeat the whole position-fixing exercise thanks to having to use a drunken Schmuck as his data recording assistant in my absence, I concluded Tom must have been joking about his extramarital exploits.

He wasn't... as I was to discover several weeks later on passing through base camp en route to Addis for my second, well overdue, spot of leave.

Finding the plane to Addis delayed until the next day, I stashed my leave bag in one of the camp's accommodation trailers and headed into town for a well-deserved evening of rehydration. Stepping through the door of the crew's favoured hang-out, the first thing I saw was a rather wobbly Tom perched on a bar stool with a beer bottle in one hand and a dusky female breast in the other.

'Ah. Two Brains. Your expert input if you'd be so kind. What d'you think of that? Perfectly normal is my diagnosis but always open to a second opinion.'

As he was talking he released his hold on his far from bashful 'patient' and directed my hand to where his had been. 'Feel anything untoward? No? Good. Confirms my reading of the matter. Next!'

Moments later, he was repeating the examination on another 'patient', one of several local 'ladies' who, it transpired, had somehow gained the impression he was a visiting dignitary with medical knowledge.

Not wishing to disappoint them, he'd asked them to form an orderly line and prepare to undergo an initial examination. Anything found to be amiss and he'd be advising a visit to their regular doctor. Up to the point of my arrival, he'd 'examined' three such patients, had declared them all normal and several more awaited his attention.

And mine. Not prepared to take no for an answer, Tom demanded I remain at his side to offer further second opinions and despite my reluctance to be a participant in this most outrageous of misrepresentation outrages, by the time I wobbled back to base with Tom and a few others, more breasts had passed through my hands than the total count in my life to date.

Needing to be up early to catch my Addis flight, I retired to my trailer leaving Tom & Co in the base camp bar discussing their own medical conditions and slept soundly in the relative luxury of a base camp bunk until roused by a disturbance in the early hours.

One glance out of the window was enough to convince me that all that talk of mistresses being essential in the foundation of a solid marriage wasn't the jest I'd taken it to be. The disturbance was sourced to the sight of a bollock naked Tom charging down the dimly-lit track between cabins yelling for someone to stop the damned bitch who'd just absconded with his wallet before going headlong in the dust, the wallet-snatcher racing away grinning broadly in the knowledge that she'd been saved by the laces of her pursuer's untied boots.

Sleep did not come easily after that, the image of the incident loop-taping in my head until I wasn't sure whether I'd actually witnessed it or dreamt it.

The answer came as I boarded the plane the next day, not the still grounded DC3 and not one that was going to Addis direct. Blott had ordered the Porter's larger sister aircraft, a twin-prop Dornier, to take those going on leave to Gode some distance away for transfer to Addis on a DC6 which needed a longer runway than Kebri Dehar's.

With just the three of us going on leave I wondered why we weren't using the Porter. That was more than big enough for our small complement of bush escapees.

On ducking through the Dornier door all became clear. Some rather familiar faces were already on board, every one of them belonging to the 'patients' Major Tom and I had 'treated' the day before and who'd obviously overnighted in various other trailers so as not to miss the early morning flight. The Dornier, it would appear, had been 'leased' from the company by Blott's HookAir operation to swap one batch of 'patients' for another and transport the fresh blood to Kebri Dehar for 'treatment' at the 'surgery' operated by Major Tom & Co.

***

As shocking as this sequence of events would've been to the Two Brains of three months earlier, these days it was received as situation normal. In just twelve weeks a near vertical learning curve had taken me from wide-eyed greenhorn rookie to semi-seasoned old Africa hand, inured to such things being par for the course on this continent. The bigger surprise would have been Blott not taking advantage of both the situation and the regulation lowly position of women in Africa. This was the African way and no one seemed to turn a hair at it, including, it seemed to me, all those African women who'd found themselves with no alternative but to rent their bodies out to survive.

It was an impression that stayed with me right up to the moment I landed in Addis for my second week of leave. In the taxi to the Hotel Afrique I began picking up a vibe not felt before. Things seemed more tense, even with the normally convivial taxi drivers. Unusually uncommunicative, mine simply drove me to the hotel, took the fare and drove off without offering a smile in return for the tip. It was as if he had his mind on other things. Something troubling.

It took a couple of days to discover what. A couple of days in which I sought out an acceptable alternative to both The Cottage and the bar from which I'd been as good as kidnapped.

It was the Viking who'd suggested it, a taciturn Danish giant engaged by the Caterpillar company to look after their bulldozers in the field who'd almost killed me with an insane piece of maniac bush driving on the way up the line to fix our dozer.

By way of recompense he'd offered to keep my glass filled next time our paths crossed and I duly tracked him down to the bar he'd mentioned to claim my prize.

Two beers in and we were joined by an American I didn't know but who was clearly not unknown to the Viking. He'd got drift of what we'd been discussing – the change in atmosphere in Addis – and after first working the conversation round to my own job criteria, felt inclined to offer his own thoughts on the matter.

Based in the Ethiopian capital for an American oilfield supply company, the man agreed the city's usually relaxed feel had disappeared in recent weeks. For a start there was the hooker strike for better pay, a move that not only had me revising my beliefs concerning the degree of passive amenability I'd thought was endemic amongst the women of Ethiopia but went some way to explaining the tension I'd noticed. The men of Addis would be far from amused by the strike.

While that on its own would explain the renewed tension, said the American, the hooker strike was far from the complete story. Word was there was also unrest brewing in the military, a level of discontent that had led to an increase in their numbers on the streets, mostly on the street corners left unoccupied by the ranks of Addis hookers.

Now he came to mention it, I had noticed a bit of an increase. But military on the streets of African capitals was situation normal wasn't it? Fellow crew members who'd been round the houses in Africa had mentioned it.

'Yeah, but this is different,' said the American. 'A lot more than usual right now. Looks like they're expecting trouble and I've even heard talk of a curfew being in the offing.'

The reason, thought our friend, lay in Emperor Haile Selassie's lack of help to those suffering the effects of the scathing drought gripping the country. His lack of action to avert famine had led to serious unrest in sections of the army and Selassie had apparently mobilised the palace guard to nip any insurrection in the bud. If that didn't work, opined our friend, Ethiopia could be on the brink of revolution... just what Somalia would be waiting for. If civil war did break out, Somalia's army would undoubtedly take advantage of it to sweep across the border into disputed areas like the one the Viking and I were working in.

As the implications of that sank in I looked to the Viking for a reaction. We both knew what a Somali invasion would mean. We'd be caught in the middle between opposing forces who'd be unlikely to make much of an effort to orientate their fighting around us. In fact, we could well find ourselves being shelled from both sides, each mistaking us for part of the opposition.

Should I be worrying, my eyes asked the Viking? We'd both heard such alarmist bar room speculation before, most of which turned out to be the product of overactive booze-fuelled imaginations. In this case, coming from a man who described himself as just a cog in the wheel of a foreign company based here, it seemed likely this worrying scenario could be taken as just that and we might be well-advised to treat his analysis with a hefty dose of salt.

'Wouldn't be too sure of that,' said the Viking once the man had gone to join another group of people across the bar. 'Word is he's a bit more than just a cog in a company wheel. More like the wheel itself and running on rails that start in Washington, if you get my drift. Gotta lot to lose strategically in this part of the world, those boys, so they keep a close eye on what's going down round here.

'That's why he was pumping you about your job. Anyone operating in the field could be a useful source of information and now he knows you carry out regular recon flights towards the border I wouldn't be surprised if what you told him finds its way into the files of some crewcut operative working out of Langley.

'If anyone knows what's going on round here it's that shady lot and if they think the area's on the verge of exploding, we could do a lot worse than listen.'

Anyone else from the crew talking and that warning would have been received with narrowed eyes. Swallowing whole what any of them said and you could end up convinced snakes lived under every bed and every one of them could kill just by looking at you.

But this was the Viking talking, a less imaginative, more phlegmatic field operative you'd have trouble finding anywhere. What he said he either knew for a fact or had good reason to believe so on this occasion I wasn't about to file this particular alert under 'just shootin' the breeze'.

Even so it did still beg a question. If we really were sitting on a powder keg, did our respective companies have contingency plans for evacuating us if the fuse was lit? They must have… mustn't they?

'Wouldn't bank on it,' said the Viking. 'I've been on contract to your lot for over a year now and a company less likely to have an evacuation plan I've yet to come across. Real bunch of cowboys you're working for. So my advice is to have your own plan. I don't need one as my time's up in a couple of weeks. But you've still got months to go.'

Shuffling back to my hotel as night fell and cut-throats seemed to lurk in every dark corner I reached a decision. The Viking was right. If the default position was that 3S couldn't be trusted to have its crews' best interests at heart, it was definitely time to start building my own lifeboat. One that was both invisible and unsinkable and by the time I got back to camp I had the blueprint for just such a vessel lodged firmly in my head.

Chapter 2.7

When to launch though? That was the question. The timing had to be exactly right. Too soon and I'd be returning to a still slump-crippled UK short of the funds needed to weather the storm. Too late and I might not be returning at all. What this needed was another reconnaissance mission over 'enemy' territory to give me a better idea of the degree of urgency needed in issuing the abandon ship order to myself.

With nothing but dust devils apparently stirring on the southern front I decided it was probably safe to leave it another couple of months… while keeping a close eye on my Somali survey team. Any build-up of unusually intense chatter amongst a group who'd undoubtedly know how close we were to getting caught in a Somali army invasion and the radio waves would be buzzing with requests for another recon flight.

One more work/leave rotation went by, then another and another and with no sign of any more restlessness than usual amongst the Somali workforce it seemed I'd be having nothing to 'report' back to the Viking's spooky friend. Well, nothing other than the definite signs of growing tension noticed when on leave stints in Addis. But he'd be fully across that, complying as he had to with the newly-introduced citywide curfew.

In theory, we all had to comply with it. But coming into effect at a time of night that allowed for serious lubrication before the due cut-off time, if the army and police had to round up all curfew breakers on the streets the jails would be bursting at the seams. So only rarely was anyone even reminded it was past their bedtime and, with the hookers firmly back on the streets and bar stools, Addis nightlife continued to thrive as if no curfew order had ever been issued.

Or it did until mid-1974 when things started hotting up.

For the crew, the first sign that the authorities were getting serious about it came the night a few of the boys out and about in Kebri Dehar for an oil change found themselves being herded back to base camp by the police. The curfew, it seemed, had spread far beyond the Ethiopian capital and this time they meant business... unlike the crew's Kebri Dehar female companionship for whom, to their fiercely vocal disgust, it meant quite the reverse.

When word reached me out in the field that security was being tightened, I knew the time had come to initiate Operation Lifeboat. The others could hang around if they liked. But for me, getting as far away from Ethiopia as soon as possible became priority number one and I knew exactly how to go about it. No one would want someone around showing all the signs of a hepatitis infection.

The only problem lay in how to fake the condition. From what I'd heard the signs were unmistakeable, notably having no control over one's bowels. Could I really get away with pretending to be caught short twenty times a day? With no proper latrine at any field camp, our 'dumping' ground at every one was on the other side of the airstrip, downwind from the camp. But since it was only obscured from sight by a few clumps of flimsy drought-ravaged thorn bushes any pretence at being in the grip of Montezuma's revenge would soon be undone.

For several weeks how to crack the problem plagued me. No sign of having contracted hepatitis, no plan.

With no possibility of fooling Big Foot in particular I was about to give up on Plan A and start wracking my brains for an alternative when serendipity fortuitously intervened. Not the sort I'd normally welcome but providential under the circumstances. After a few more stints in the field I really did get the shits... caused by having to resort to drinking water from the water bag.

For the umpteenth time since my arrival the regular supply of bottled water failed to arrive leaving us having to dig into the personal reserve stashes every crew member salted away for times like these. Under normal circumstances they were enough to keep us going until Blott got off his indolent butt and made sure fresh supplies were on the next plane.

Not this time, and with my reserve the first to run out I was left having to beg for sips from the others' supplies. When theirs started running low as well, personal survival instinct kicked in, all pleas for the odd drop to avoid expiring from dehydration were refused and I was left no option but to get better acquainted with the water bag.

The problem wasn't contamination. Once boiled the water pumped up from Digit's wells was fine. But no amount of boiling could rid it of the slightly salty taste that came with water drawn from magnesium-tainted subterranean reservoirs.

While the constitutions of our Ethiopian and Somali crew members were well used to it, for the uninitiated the effect was explosive. In moments one's innards would begin twisting and gyrating and sending the drinker in urgent search of a toilet roll and somewhere to poison both the ground and the atmosphere. Epsom salts had nothing on Ogaden well water and it once occurred to me that should we not find oil, to recoup some of its exploration losses the company which had sent us in search of it could do a lot worse than bottle the stuff and sell it to health kick junkies on the street corners of Los Angeles and New York.

So unless one really HAD to purge the system, the water bag water was reserved strictly for washing and filling vehicle radiators. Drink it and you knew you were entering vicious circle territory. Starting with dehydration through diarrhoeatic loss of bodily fluids, the cycle would move on to raging thirst and then to gradually escalating attacks of more diarrhoea and more dehydration as one tried to quench it with gulps of yet more water.

That I was getting into dangerous territory I knew full well, once even resorting to brushing my teeth with gin to avoid ingesting well water by mistake. But now I had no choice and the result was as predicted... with one exception. It almost led to my demise by aeroplane propeller.

Two full days on water bag water and the airstrip was getting as used to my regular sprints across it as I was in forecasting when the sprint needed to start. On every occasion except once my forecasting skills proved accurate. But it was a once I'd remember for a very long time and for all the wrong reasons.

With the world threatening to fall out of my bottom and shorts round my ankles in the middle of the airstrip I presumed I was safe. Protocol required that any plane wanting to land should first do a warning pass over the camp and there'd been no fly-by at the time I set off across the strip.

So it came as something of a shock to see the Porter swooping without warning up over the thorn tree tops to begin its landing approach to an airstrip on which a member of the crew was, in effect, shackled by his shorts.

Just one thing went through my mind. At what point would I have to abort my own mission and dive between the Porter's wheels to avoid a real life re-enactment of that horrific scene in 'Catch-22' when McWatt accidentally slices Kid Samson in half with his propeller. The timing would be crucial.

As the Porter closed in I braced for that life-saving leap and started praying that Fishball – for this HAD to be him – would spot the arm-flailing semi-naked man with diarrhoea streaming down his leg in time to abort the landing. Only one of us had the power to avert what was looking like becoming a nasty encounter with a bacon slicer and I prayed that Fishball was having one of his more observant days.

At first it seemed my prayers were being ignored. Closer and closer the plane got without deviating. Then, at the last moment, it dipped sharply to hit the airstrip hard and bounce the aircraft over me, wheels just missing my head.

It returned to Earth just beyond, slewing on down the airstrip as whoever had performed this bunny hop landing manoeuvre fought to bring the plane under control.

Watching astounded as it came to a halt in a cloud of dust inches before careering into the end-of-airstrip bush, I realised that my own original mission was still far from over. A mission that required an onward shuffle to the bushes beyond the airstrip, one hand holding up heavily soiled shorts, the other making obscene gestures to the maniac Fishball for yet again ignoring the landing protocol.

I didn't need the water bag water to help me with my second bowel purging in as many minutes. Fishball's antics had proved a more than adequate substitute and after finishing what I'd come to do I stormed back to the mess tent to inform him of the fact.

Detouring to the shower en route to cleanse myself with the water that was responsible for my condition in the first place, I tried to compose both myself and the case for the prosecution. Besides reminding Fishball of the protocol, the plan was to get it on record that all this could have been avoided had the base camp 'management' and its re-supply transport team bothered to act promptly to replenish our stocks of drinking water when requested to do so.

Ducking into the tent quivering with rage and ready to give the man a seething mouthful of incandescent fury I froze in mid-stride and mid-rant. Sitting at the table sipping from a newly-opened bottle of Evian water was the Red Baron, looking straight at me with fire projecting from his eyes.

'Motherfuckin' shitferbrains limey!' he exploded. 'Ya tryin' ter fuckin' kill me? If yous gotta death wish that's your affair. But do it in ya own time, asshole, and don't go tryin' to take others with ya!'

Already drained from dehydration I felt whatever blood was left in my face sink to my stomach and start churning. Then a lava flow of bile started rising up my gullet and before I could stop myself grabbed the bottle from the Baron's hands and upended it into my parched mouth.

Big mistake. Too late I realised the water in the bottle was of the sparkling variety. Reacting violently with the magnesium residue in my stomach an effervescent mix of water and the remains of my breakfast exploded back up the pipe with such force I was unable to stop a pressurised gusher of rancid primordial soup projecting clear across the table to land with some accuracy in the Baron's lap.

Now it wasn't just me lost for words. In their place, dinner plate eyes conveyed volumes as the Baron sprang from his chair dripping vomit and Southern State fury to lurch hysterically over the table to make a grab for my throat.

Too weak to react, I just stood with my knees quaking before they gave way altogether and I collapsed into the nearest chair leaving the Baron's hands grabbing at thin air.

'Whoa. Hold on horse,' I heard someone say. A voice I didn't recognise was intervening to bring what passed for decorum amongst the crew back to the proceedings. 'Lay off him, Joe. The boy's clearly sick. Look at his face!'

Revived marginally courtesy of small sips of non-sparkling water from the consignment the Baron was ironically delivering, all was gradually revealed to me. Not trusted to ferry one of our client's visiting wheels out to the field to check on survey work progress, Fishball had been left nursing his ego at base camp while the Baron did the chauffeuring and I had the wheel to thank for saving me from the infuriated chief pilot, now fled to the shower I'd recently vacated to hose himself down.

'Take it easy son. Slowly,' was the drinking instruction being murmured in my ear by a man my blurred vision detected seemed to have silver tips to everything. From pointy cowboy boots to pointy cowboy collar and pointy bootlace tie to pointy trouser belt, every point that could be silver-tipped had been and that, I decided, went for the silver-haired stranger's tongue too. His calm, authoritative order for the warring factions to desist had defused the situation in seconds and left no one in any doubt as to who was in charge here.

By the time the Baron returned from his dousing down, the wheel had not only brought calm back to the proceedings but had assessed the situation and made a decision. 'The boy', he told a still fuming Baron, needed medical help tout suite and would therefore be accompanying them back to base to see The Gravedigger.

I could hardly believe my luck. Here was a gilt-edged opportunity to put Operation Lifeboat into effect and I had one of the company's client's most influential operatives to thank for it. There on a plate he'd handed me the chance to convince the company medic I'd been struck down by the dreaded hepatitis and, knowing the medic concerned, I was one hundred percent confident he could hardly diagnose anything else. How could he after having been recently informed that his secret 'arrangement' with Blott's wife wasn't as secret as he thought?

Chapter 2.8

To cut a long and quite convoluted story short, I escaped the revolution by a whisker. Just weeks after landing back in the UK with my phoney hepatitis certificate, Ethiopia's regular army deposed Emperor Haile Selassie and the Somalis, as forecast, had rampaged across the Ogaden border forcing the 3S seismic crews to abandon ship and run for their lives.

Listening to news of the insurrection from the warmth of Flame Hair's bed, I had trouble suppressing a smile. Had I not looked out of that hotel window at that late night moment during that period of leave from the field I'd undoubtedly have been running alongside them. But I had and that gave me all the ammunition I needed to assure myself of The Gravedigger's full cooperation when it came to launching my pre-revolution hepatitis-inspired self-evacuation plan. Refuse me an infection-positive certificate and word might get out that while Blott was still in the bush, the devious medic had been spotted getting out of a taxi with Blott's nearest and dearest to furtively enter the hotel opposite mine, nervously eyeing the street left and right.

On the plane back to the field I'd casually let him know he should have looked up as well. While the man was incompetent he wasn't stupid and it wouldn't take him long to work out where that left him – forever in my debt if he wanted to retain both his cushy little number and his own skin.

On such small moments do fortunes turn, I thought as the Baron piloted the Porter back to base camp with me on it.

With that IOU from my now favourite doctor in my pocket, there'd be no complications in obtaining the necessary infection certificate, a letter from him recommending immediate repatriation to the UK and subsequently both a ticket home and my end-of-contract bonus, both of which the company had every right to withhold if I simply walked off the job.

Was that a pang of guilt making a fleeting appearance as I rolled over to feel Flame Hair's naked body pulsing gently in sleep beside me?

Not exactly. If there was any guilt to be felt the only ones feeling it should be the 3S management who'd lied through their teeth to entice me into their pernicious embrace. Almost nothing they'd told me had been true and had this been the US there'd be a string of ambulance-chasing lawyers at my door begging to be engaged to launch a breach of contract law suit.

But this wasn't the US and the company not only knew it but had taken full advantage of it. Here in the UK, 3S was not unaware that anyone signing on with them effectively lost all worker rights the moment their pen hit the paper. UK legislation designed to protect an employee's interests did not extend to British citizens working for foreign companies overseas and while this scumbag operation knew it full well, they'd been very careful to avoid making any mention of it in their recruitment patter.

Well now they weren't the only ones wised-up to the loophole and it was something I'd be keeping well in mind when it came to casting around for new employment. With a pocketful of cash built up over the past few months, there was no urgency to get back on the horse. I could bide my time over re-mounting it. But sooner or later the well would start running dry and a decision would have to be made. Hunker down in the UK or cast around for a way to return to the seismic fold.

For the moment though, such decision-making could wait. I was back on home turf and feeling the need to celebrate. With confirmation just received that I hadn't been struck down by hepatitis, friends were gathered to drink the health of a man who, though he'd never know it, had been instrumental in getting me out of that godforsaken place. Had Haile Selassie not been such a shit to his own people the revolution would never have happened, there'd have been less motivation to launch Operation Lifeboat and I'd probably still be stuck out in the back of beyond wondering how I was going to extract myself from this self-inflicted purgatory. The very thought of it had me going weak at the knees.

***

Of all the questions friends had on being appraised of the events of the previous few months, the most common was 'what now?' After being so comprehensively shat upon after being led by the nose up the garden path, would I feel inclined to risk it again?

Looking around at what Britain had become in my absence – a grey, lifeless, depressive place devoid of any reason to celebrate anything – the answer was so obvious, and so potentially upsetting to those around me who couldn't see what I was seeing, that I gave it only to myself: what choice do I have? Stay and within months I'd be indistinguishable from all those I saw sinking ever deeper into the mire of economic gloom and despondency. Overseas, at least I'd be spared that.

So, as much as I tried to give the impression of not being sure, deep down I knew the answer. Pretty soon I'd be on my way again… providing my history to date in the claustrophobic little world of seismic survey work hadn't made me *persona non grata* to the industry. With everyone in this world knowing everyone else there was every possibility I'd ironically burnt my boats with the launch of Operation Lifeboat. Word could well have got around about how I engineered my flight out of Ethiopia and if it had, my name could already be on a not-to-be-trusted blacklist.

Surely not, I tried to reassure myself. Surely, in a world full of sharks pulling stunts that made mine look like a minnow's flight from danger I'd surely have to try a lot harder if I wanted to attract the attention of the man with the blacklist pen. The industry was full of far bigger fish for him to fry.

Even so, the worry persisted. Would I find every door slammed in my face when I got around to knocking on them?

The answer came when I opened my own a few weeks after my return. There outside was a presence that confirmed my darkest, most buried suspicions – that while you could run from the oil industry, remaining hidden from it and beyond the reach of its vindictive omnipotent tentacles was a different matter altogether.

***

'Hello, Judas,' said Whackamole to my drop-jawed face. 'How's the hepatitis?'

Judas? Oh, right. The crew must not only have put two and two together but to have jointly felt a bit miffed by my little repatriation ploy. Fair enough. I'd probably have felt the same. But enough to warrant the dispatch of Whackamole and Big Foot to my door to make me aware of the crew's feelings? I was hardly just around the corner from their respective homelands.

On the other hand, maybe they'd made such a massive detour from their trips home to congratulate me on my ruse. Knowing my fellow crew members like I now did that wouldn't surprise me one bit. Such was their general character they could well have reacted with some admiration over finding one of their number so devious he'd found the means of extracting himself from the purgatory of the field without fear of losing his end-of-contract bonus and had put his plan into effect without telling them.

If that was the reason for their visit they had a funny way of showing it… as soon became evident once they'd pushed unceremoniously past me to begin liberating the alcoholic contents of my fridge.

Slumped in my favourite living room chair, booted feet planted solidly on the coffee table, Big Foot began recounting the events of the past few days, steely eyes fixed firmly on me to gauge my reaction.

As the country descended into chaos, he said, the crew had had to dodge bullets to get out and he hadn't been the only one envying me for having managed to avoid this little bit of unexpected distraction. That envy, he said, had turned to suspicion the more the crew members thought about it. Wasn't it just a bit coincidental that I'd got out just before the storm broke? Did I know something they didn't and had kept it to myself?

If I did and had, Big Foot scowled, I shouldn't be surprised to find a posse on my trail if not a baying lynch mob lusting after my blood. In the doodlebugging world – the nickname for all things seismological and geophysical – the done thing was not to let your 'brothers' down, he snarled. Them that did risked finding scorpions in their boots and snakes in their beds the next time they were in the field.

There was only one way to avoid that, he said. Since I was new to the game, if I HAD engineered a vanishing act and kept it to myself they'd let me off lightly this once providing I fessed up and offered an unreserved apology. 'Leniency for those who confess,' he growled into his beer quoting the Chinese cultural revolution system of mob justice at me, 'severity for those who deny their crimes.'

Just how lenient their idea of leniency might be I had trouble telling from Big Foot's face and he seemed in no hurry to enlighten me. But of one thing I had no doubt. If I caved in and confirmed their suspicions, no matter how much they tried to assure me I'd be safe from retribution there'd still be a risk to health. My financial health.

The problem was that, by simply showing up at my door, Big Foot and Whackamole had proved beyond doubt they were to be trusted about as much as my erstwhile employer. No one other than the 3S recruiter in London knew my address and there was only one way the pair could have got hold of it. All of which meant their claim to have tracked me down to give me a chance of avoiding being ostracised by the doodlebugging brotherhood was about as genuine as my hepatitis infection. The moment I found them outside my front door I knew the real reason for their visit. Still clearly attached to 3S, they were here on behalf of the company to extract a confession that'd give them clear grounds for rescinding my end-of-contract bonus.

That, I'd decided after all of a microsecond's thought, was not going to happen. Retaining that bonus wasn't just desirable, it was vital to my future employment prospects. It'd buy time to check future employers out properly and hopefully avoid finding I'd signed on with another cowboy outfit that thought nothing of dispatching its employees to places on the brink of violent insurrection.

So no, there'd be no tearful shamefaced confession from me on this occasion nor any appeal to my 'brothers' to go easy on me. The only thing Big Foot and Whackamole would be leaving with was the picture in their mind's eye of my look of shocked astounded innocence and indignant offence over them even THINKING I might've had an ulterior motive. I hadn't needed a revolution to put my life at risk, the look went on to say. The dreaded hepatitis was already doing that and had I not left when I did, there was every chance of me being shipped home in a box regardless of whatever else was going down in Ethiopia. So, the look concluded in saying, you can just take your 'Judas' accusation and shove it!

For a moment Big Foot and Whackamole just looked at me, scanning my features for signs of bluff. They never saw it. To my own amazement I managed to hold firm and in the end it was them who blinked first, appearing satisfied with my lip-quivering indignation and eventually giving up and sloping off back to their hire car to make it to the airport in time for their onward flights.

Having topped my performance off with a peevish refusal to shake hands as I showed them the door, I watched unsmiling as they went, holding my breath until they were well out of sight.

My God, I thought, disappearing back inside to pour myself a stiff drink. That was close. For the moment I reckon I'm in the clear. But stay on your guard, young man. Who's to say they weren't just the forward expeditionary force sent to sniff the air?

Knowing the oil world as I now did, that possibility could be far from discounted. There was every chance of them reporting their findings back to 'someone' higher up and of that someone deciding to follow it up with a full-on invasion, company troops sweeping over the hill to take up positions outside my front door and remain there until I capitulated. Fired up with face-saving zeal to avoid becoming the laughing stock of the oil industry, I could be fairly sure that if that was the eventual outcome, the recrimination would be such it'd leave me in no doubt that trying to put one over on those who considered themselves impervious to any Earthly power was some distance from the brightest idea I'd ever had.

Chapter 2.9

'Well bless the star in my hairy old fundament!' a voice I recognised was exclaiming from the seat behind mine. 'If it isn't young Two Brains! Haven't seen you since you were laid low with that nasty infection. How's that coming along?'

Looking round to confirm the speaker's identity I was met with several slightly open-mouthed glances from fellow bus passengers before my eyes came angrily to rest on the infection-mentioner. I was right. There, not twenty-four inches from me, was the man I'd spent a long and very voluble night with under the stars.

'Chrissake, Tom,' I hissed. 'Keep it down will you… unless you've no objection to others hearing how you once masqueraded as a doctor specialising in mammary conditions!'

Now the ears as well as the eyes of all on the bus's top deck were on us, every one of them desperately hoping keeping it down was the last thing he was going to do.

They were to be disappointed.

'Ah. Point taken, old chap' he whispered, shifting his bulk to the vacant seat beside me and crushing me up against the bus window. 'Perhaps that was a tad indiscreet. One forgets one is back in polite society sometimes. Blame it on that extra shandy I had at the cricket. Damn fine game. You there?'

Although I'd have liked to have been I'd had other things to do that afternoon but had no intention of going into detail with him about it. For all I knew Major Tom might be part of the follow-on softening-up force I feared, dispatched by 3S to use his genial skills to get me to drop my guard over the hepatitis scam through engaging in idle chitchat about my movements. So instead, all he got was a blank look.

'Not a cricket fan?' he said. 'So sorry. Thought you mentioned you were last time we met. But that might've been me. The night sky leaves me a bit distracted at times. My lovely lady wife's always saying so. Alpha Centauri-eyed, she calls it.'

'Yeah, I can see how that could happen,' I said before I could stop myself. Damn. Now he'd got me talking and if he was here as a company stooge there'd be every chance of me not spotting his googly to find I'd dropped myself in it without realising. So be very careful to engage brain before opening mouth, I scolded myself. Could be the difference between staying this side of solvent and having to reacquaint myself with the benefits office.

'Thought you might,' said Tom. 'You were one of the few assistants I've had in which I saw a bit of myself. Not everyone gets bedazzled by the stars. But you did... unlike that pisshead Schmuck. Only thing he seemed interested in was the bottle in his hand and preaching the virtues of the apartheid system at me, horrible man. He was the reason you and I had to do that extra astro stint. Couldn't make head or tail of the readings he recorded the night before so had to do them all over again. You can't believe how relieved I was when he got on the plane and someone blessed with a bit of grey matter got off.'

'Me?'

'Sure. You're wasted as a bum line surveyor, you know. You should get trained up to do proper survey work. Might stretch you a bit, maths-wise, but more than worth it. You wouldn't regret it.'

'But I AM trained up. Even managed to put some of it into practice before the economy went tits up and I had to take the line surveyor gig to survive.'

'Really? Then you, young Two Brains, might be just the chap I'm looking for. Got a few astro gigs coming up over the winter and I could use a reliable assistant. Fancy that? Please say yes. You'd be saving me from a fate worse than Schmuck.'

***

With massive restraint I remembered what I'd told myself about engaging brain and didn't immediately bite his hand off. This could easily be a trap, I thought. Part of a 3S strategy to get me to confide in Major Tom about the hepatitis thing. He'd no doubt want assurance I was clear of it before taking me on and that'd mean having to give him sight of the certificate confirming I was hepatitis negative… and always had been.

Anyway, that apart there were one or two things about our Tom and his job offer that didn't quite add up, not least his chosen mode of transport. Watching him alight from the bus and wend his unsteady way into the railway station to catch his train home, it occurred to me that if his business was going as well as he claimed, what the hell was he doing going to the station on a bus? It'd be something I'd be alluding to at the meeting he'd proposed to avail me of the astro gig details.

'Bad day at the races,' he explained when I went to check the job offer hadn't been just the 'shandy' talking. 'Backed a few bad 'uns at the cricket so the cupboard was a trifle denuded by the time I had to leave to get my train. Damned barman wouldn't cash a cheque for a visiting team supporter so the jolly old omnibus it had to be. But clouds and linings eh, old bean. If I'd had the wherewithal for a taxi our paths wouldn't have re-crossed would they?'

'So business is doing OK then…'

'Never been better. With the oil price the way it is there isn't a seismic crew you can name that isn't out there carving lines through the bundu, every one of 'em wondering where the fuck they are and getting on the blower for an astro fix sharpish.

'But don't blame you for asking,' he added. 'I'd also be being a bit careful about who I threw my lot in with if I'd been through what you went through. Not the easiest of bloodings into the oil-hunting game from what I've heard.'

OK, I thought. Here it comes. This is where he begins pumping me for details about the hepatitis and my great escape from Ethiopia so keep your wits about you.

Had he read my thoughts or was he really on the level? For rather than start pulling at the threads of my time with 3S and quizzing me for details of my departure from the field he pulled out a thick folder and shoved it under my nose.

'There you go, young buck. Think that'll set your mind at rest. That's the incoming astro assignment request reject file. Either cos I didn't fancy the location or the timing or who the request was from. There's some decidedly dodgy folk out there… but you already know that. I only took the Ethiopia gig cos another job fell through at the last minute and I found myself temporarily short of the green folding stuff. Otherwise that job offer would have ended up in the folder you're holding. Right bunch of shysters you were working for if you don't mind me saying so.'

Far from it, I thought but didn't say. Join in the 3S rubbishing before I had a much better idea of who I was talking to and I could well find myself trapped into spilling the great escape beans without realising it. Although the end-of-contract bonus amount was so piffling it wouldn't even have registered in the company's petty cash accounts, I knew they wouldn't hesitate to use everything – and everyone – at their disposal to retrieve it. There was a principle at stake here. About the only one my former employer had. So I'd need a bit more evidence that Major Tom wasn't still in 3S's pocket before he'd be getting any indication of there being more to my repatriation than a genuine medical emergency.

***

Leafing through the folder Tom had handed me it gradually dawned that I had no real need to worry about him. There amongst the declined offers of astrofix work were three requests from my former employer and three responses back to them, every one written in such a way as to leave them in no doubt that his experience in Ethiopia had left as much of a mark on him as it had on me. Every letter to 3S was crafted in language designed to leave them in no doubt that he only worked for organisations he considered reputable.

Well that seemed to settle it. Major Tom could (probably) now be trusted and, in the light of any better offers, joining him on his upcoming assignments seemed a generally risk free employment option providing no formal strings were attached. The last thing I wanted was to find myself having to engage a raft of lawyers to extract myself from the arrangement should there be any sort of falling out between us – always a possibility when stuck out in the back of beyond with someone you'd had few prior dealings with. From previous experience we both knew that.

Anyway, by keeping things on a strictly informal footing I'd feel no obligation to give him chapter and verse about my medical history, the hepatitis thing in particular. Despite everything, I still felt the need to keep that under my hat. As Tom had said, there were some 'decidedly dodgy folk' out there and the fewer people who knew about my little ruse the better. In one of his more 'relaxed' moments, he might let slip about it in mixed 'dodgy' company.

So as we geared-up for the first of the assignments – out in the badlands of northern Kenya – all talk was restricted to cricket, what we could expect when out in the field and what precautions might be necessary to avoid nasty surprises catching us unawares. By its very nature, seismic survey work had to be carried out in remote locations where the natives might not be overly enamoured by our presence and it was always best to go into the field prepared to expect the unexpected.

As things turned out, in the five astro-fixing assignments Tom and I carried out together in various corners of Kenya, Tanzania, Uganda, Sudan and Somalia in succeeding months, not once did we fall seriously out and only on the rare occasion did we encounter excessive hostility towards our presence. Tom's generally affable disposition ensured that.

But then the northern winter came to an end, Tom shelved all overseas travel while his beloved cricket was being played in England and I was left with a difficult decision to make. I could either keep dodging my bank manager during what would be a financially lean summer or start casting round for alternative employment. Unlike Tom, not only had I yet to build a decent tiding over fund but under the man's wayward guidance had actually seen my own reserve shrinking. To such an extent in fact that, once again, I found myself in the position of having to count the pennies.

Tom's advice over 'shandies' at his local cricket club? Seasonal work. Maybe as a pro cricketer? That'd help keep the bailiffs from the door (hic) and would allow me to make myself available for another winter of astro work when the time came (belch).

After considering his advice for all of five seconds I decided to follow my own. As enjoyable as he was to work with, I – or more precisely my bank manager – couldn't afford him, with or without the sort of employment any pro cricket club was unlikely to offer me. So as much as I regretted having to say it, there was little option but to return to year zero and start scanning the newspapers for survey work job offers.

This time, with every oil company in the world desperate to cash in on the soaring price of oil, I could take my pick of exploration survey jobs advertised and in the end I settled on a company we'll call, for the sake of saving the blushes of innocent parties involved, Off The Charts Ltd, a London-based survey outfit with what looked like a rather better track record than my previous employer. Being British, at least it'd be less likely to try flouting UK employment law than their US counterparts.

Two days after phoning to give a summary of my career to date I was in the survey manager's office signing an employment contract and within the week I was back on a plane to Africa. So desperate were they for people with the right skills no request had been made for references, the 'interview' had been as cursory as the follow-up medical and from the distracted look on the manager's face throughout the interview I got the distinct impression that his only consideration was making sure every unoccupied hole had a peg in it, regardless of the shape of either.

My suspicion was confirmed on receiving the OTC job confirmation letter a couple of days after the 'interview'. With complete disregard for all my talk of having no field experience beyond semi-arid regions, the manager had listed me for a job in the jungle. Not just any old jungle but one of the densest, tangliest, swampiest, most inhospitable, disease-ridden jungles in the world. The one covering almost the whole of equatorial West Africa that came complete with every venomous biting creature known to man and the thoroughly appropriate Victorian nickname of the White Man's Grave.

Part Three

GABON 1975

It's a jungle out there

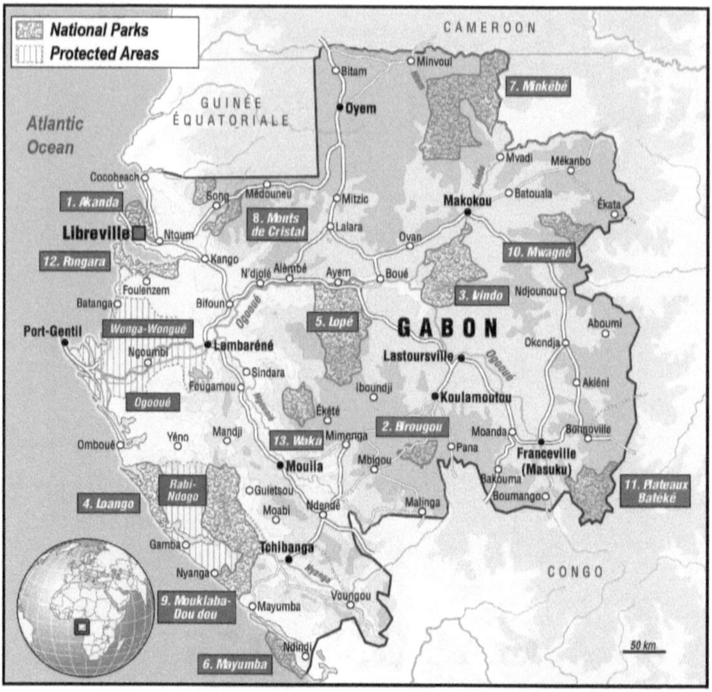
National Parks
Protected Areas
CAMEROON
GUINÉE ÉQUATORIALE
Atlantic Ocean
GABON
CONGO
1. Akanda
2. Birougou
3. Ivindo
4. Loango
5. Lopé
6. Mayumba
7. Minkébé
8. Monts de Cristal
9. Moukalaba-Doudou
10. Mwagné
11. Plateaux Batéké
12. Pongara
13. Waka
Wonga-Wongué
Ogooué
Rabi-Ndogo
Libreville
Port-Gentil
Lambaréné
Oyem
Makokou
Lastoursville
Koulamoutou
Franceville (Masuku)
Mouila
Tchibanga
50 km

Sindara
Fougamou
Ogooué
Ékété
Omboué
Yéno
Mandji
13. Waka
Mouila
Rabi-Ndogo
4. Loango
Guietsou
Moabi
Ndendé
Gamba
Tchibanga
Nyanga
9. Mouklaba-Dou dou
Mayumba
Ndindi
6. Mayumba

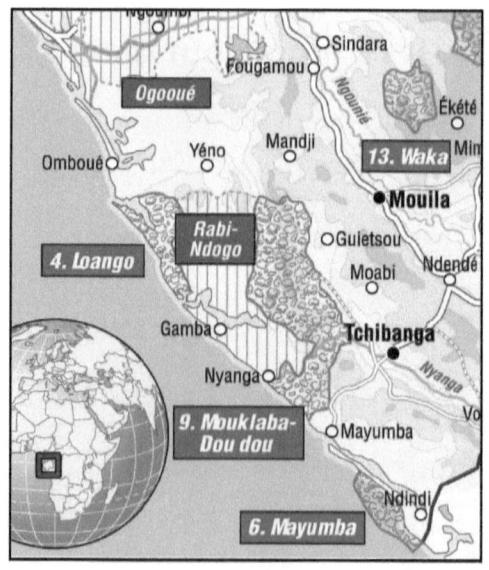

Chapter 3.1

Mesmerised by the black mamba's glittering eye as it slithered across the bonnet towards the shattered windscreen, I couldn't help thinking there must be an easier way to make a living than this. Maybe as a surveyor for my local authority? The most dangerous thing I'd have to confront there would be dodging bricks hurled by residents showing their displeasure over someone updating maps of council estate land earmarked for sale to private developers. A doddle compared to this, a little number that entailed hacking one's way through virgin tropical rainforest of mile-high trees full of deadly African bees, venomous frogs, black widow spiders and sundry other life-threatening creatures, every one of them taking great exception to having their natural habitat disturbed.

On the other hand, if I was still serious about gaining a solid toe-hold in the world of oil exploration, taking a job with a local authority might not be the best way of going about it. If the oil industry had plucked anyone from the ranks of those favouring nine-to-five council jobs for life, I'd yet to come across them.

But even if I was the first I'd probably still end up here, trapped in the cab of a Landrover pickup at the bottom of a Landrover-sized hole made invisible by elephant eye high grass. With the number of oil prospecting projects underway in the leafy pastures of rural Sussex being roughly on a par with the number of council employees drafted into the oil world, you could almost guarantee you'd end up somewhere like this. Places where there weren't many around to see what the oil companies were up to.

Or at least not many with a voice. In Africa one of the first things you learn is that no matter how remote you are from civilisation you're never quite as alone as you think and the jungles of Gabon were no exception, thank God. For had it not been for the voiceless ones appearing out of nowhere courtesy of the mysterious magic of the bush telegraph I might still be there to this day, cuddled up in the Landrover driver's seat with one of the most deadly snakes in the world.

***

The proper, tropical rainforest description of the vegetative mass completely covering Gabon hardly did it justice. As I was to discover from six solid debilitating months in it, here was the real jungle McCoy, an unadulterated tangled mess of primaeval metres-thick trees, curtains of dangling lianas and all the impenetrable undergrowth and swamps of the average Tarzan film. A veritable slew of vegetation into which I'd been plunged with no training whatsoever to map terrain impossible to photograph from the air. Having been granted an exploration concession for the area, before drilling any exploratory wells the Shell oil company needed to know what was under the forest canopy and the result was my presence here gathering data for the necessary maps, the non-existence of which meant that, so far as I and my team knew, we were the first non-Africans ever to set foot in this particular bit of Africa.

For non-Africans read sacrificial lambs. Not only was leprosy rife here – the famous Dr Schweitzer set up his renowned leper colony in Gabon – but so were any number of life-changing parasitic diseases carried in the gut of the mosquito and tsetse fly swarms that went everywhere the field worker went.

Landing silently to suck blood with such relish one could be forgiven for thinking white man's blood held special curative properties for these micro vampires, fresh meat like me were singled out for special attention and we quickly developed a protective, machete-based, technique for dealing with it. With sweat-soaked shirts a magnet for huge armour-plated tsetses in particular, on treks through the undergrowth one got used to the feel of the slap of a machete on one's back as the man behind put paid to yet another tsetse or mosquito trying to infect you with malaria, dengue fever, sleeping sickness or even the dreaded elephantiasis.

As one with an understandable horror of all of these, to ensure there was no shortage of insect splatterers behind me I made certain of always being at the front of the line, something that came as a considerable relief to those trailing behind. Not only would I be the first to step on any viper, mamba or other venomous beast too slow, deaf or vindictive to get out of the way but being the one least practised in the art of insect swatting with a razor-sharp machete the others' backs were less likely to end up looking like they'd had a close encounter with a combine harvester.

By the end of my Gabon tour my machete-wielding technique had improved and despite several near-misses no viper or mamba ever got me. But with the scars of any number of successful tsetse attacks abundantly apparent, the jury's still out on the other diseases. With body parts not obviously inflated beyond normal ranges it looks like elephantiasis has been avoided. But whenever I get a fever or have trouble getting out of bed, my automatic reaction is to blame the beasts of Gabon and the company that sent me there with virtually no preparation or protection.

***

Despite doing what research I could on OTC I still crossed my fingers when it came to signing on, hoping to hell I wasn't getting into bed with the UK's version of the lying, cheating, conniving shysters I'd worked for in Ethiopia.

So much for crossed fingers then. On arrival it became all too apparent that OTC too had its shortcomings, not wholly disconnected to the calibre of a workforce that made up the bulk of its Gabon contingent. Primarily composed of initiative-lobotomised ex-British army milliterati incapable of taking any decision without orders from above and with zero experience of working with anyone from a different background it instantly became clear that a close and harmonious working relationship between the two groups might take time to achieve.

Correction. For 'take time to achieve' read 'never', as quickly became apparent the moment the initial forty-eight hour familiarisation with my surroundings honeymoon period was over. Forty-eight hours in which my workmates-to-be lulled me into an initial false sense of security with the warmth of their welcome to OTC's R&R base, Gabon's harbourside town of Port Gentil.

Given a full thirty minutes to check into my hotel room and recover from a gruelling overnight flight from London to Port Gentil by way of Paris and the Gabonese capital Libreville, my welcoming committee of fellow jungle workers were in no mood to take prisoners in their bid to ensure I got the full flavour of the town's hospitality hotspots before I was whisked away to the survey crew's jungle base.

Or rather, they were. Taking prisoners, that is. Waving away any protest to at least let me rest up for a couple of hours, the banging on the door continued until the new inmate caved in and agreed to go quietly, collared and dragged out into the debilitating heat and humidity of the equatorial midday sun and thence to the crew's favoured watering hole, a ramshackle wharfside bar where the *moules marinière* came in buckets.

Ignoring the gasps of disgust from the waiter and fellow restaurantistes as my captors ordered beer and RED wine with theirs, the buckets kept coming and coming until the restaurant ran out and the entourage repaired to their respective rooms to recharge their batteries in preparation for stage two of the welcome to Gabon festivities – back to the Wharf Café for steak, frites and games of high stake alcoholic poker until focussing on our cards was no longer possible.

My pillow hardly knew what hit it after all that. Collapsing into bed hog-whimpering drunk, penniless and exhausted in the early hours all I could think of was that at least I was now entering the recovery period every jungle worker on leave allowed himself before starting the whole sleep/eat/drink/gamble one's hard earned cash away cycle again the next day.

Passing from semi-consciousness to full-on insensibility I hoped that'd be enough for my travel wracked body to repair itself. Although there was no jet lag as such, Gabon being roughly on the same meridian as London, the twenty hours it'd taken to get here had taken its toll and the jump from British summer drizzle to pressure cooker equatorial heat and humidity wasn't helping.

I was never to find out. The hammering on my door at six in the morning made sure of that.

'Plane leaves in an hour. Be on it,' a voice I didn't recognise was chainsawing into my head. 'No excuses.'

What? Not even travel-related paraplegia aggravated by terminal alcoholic poisoning?

Apparently not. The thumping heads poking blearily out from other hotel rooms confirmed it. The holidays were over and as of six a.m. we were back on company time.

***

Despite not having to pack the bag I'd not had time to unpack in my fellow workers rush to initiate me into the delights of PG, I was the last to arrive at the airport. With constitutions honed to the resilience of the average African goat through many such leave breaks from the jungle, by the time I emerged every one of my workmates had scampered largely breakfastless to the sheds that passed for a flight terminal, the reason soon becoming obvious. The last to arrive had no option but to take the rearmost seats of the ten-seater Islander aircraft we were boarding, the ones on the receiving end of maximum turbulence as the plane lurched its way through the tropical rainforest updraft to Shell's oilfield base at Gamba, some 250km south of PG.

Those like me who'd taken time out to grab a hunk of baguette and slurp a quick coffee en route to the taxi rank instantly regretted it the moment the plane got to cruising height above the solid green mass a few hundred feet below. Here, air pockets were the rule rather than the exception and even a four-point seatbelt wouldn't have been enough to prevent the rear seat occupants becoming intimately acquainted with the ceiling as the aircraft leapfrogged its way along the West African coast.

By the time Shell's oilfield gas flares came into view on the Gamba outskirts there was more baguette and *moules marinière* in three vomit bags than in my stomach and a communal look of relief on the face of every other passenger. At last they'd be escaping a stench that'd had them tearing in desperation at the plane's unopenable windows and reaching for their own vomit bags.

On landing, they fled down the steps, I hobbled, my legs and knees giving thanks for having to face no more than a few steps to get off this flying tumble drier. Any more and I'd have been going on another flight, down the remaining steps on my backside and somersaulting into the Landrover waiting to convey me to the OTC base camp fringing Gamba's coastal lagoons.

Having the camp location described to me the previous evening by one of Shell's oilfield engineers, into my mind's eye came a vision of near holiday camp perfection. Apparently, every resident had his own well-appointed tent, every one with an uninterrupted view across the lagoon to the sea and you could hardly move for birdlife in trees that gave ample protection from the tropical sun. Being but a short stroll to the beach one could take a relaxing dip in the surf to cool off and, with Gabon being a former French colony, the camp cuisine was French influenced, always on tap and always accompanied by an extensive range of fine wines and other beverages available from the camp bar.

By the time he finished describing the living arrangement laid on for those good enough to have agreed to come and help Shell in its endeavours, I'd begun to feel almost valued. Clearly, here was a company that put as much stock in its contractors as it did in its own staff and in my experience that put Shell in a class of its own. No other oil major I'd worked with seemed to give more than half a hoot for anyone regarded as ex-familia. If you weren't 'one of the boys', to all extents and purposes you were a non-person. A commodity not worth a spit that could be hired and fired at whim and replaced with another pulled from the bottomless pit of cannon fodder oil worker contractors.

So as I loaded my bag into the Landrover sent to meet me I felt the woes of the previous forty-eight hours evaporate and hopes rise that at last I'd found someone in this industry willing to demonstrate their appreciation for outsiders brought in to help them achieve their ends.

Then I saw the inside of the Landrover cab and all hope drained through the cavernous gaps between the vehicle's floor panels.

After the holiday camp description given by my Shell engineer friend, the last thing I expected was to be transported to the holiday camp in a vehicle devoid of all furniture and instruments save for the driver's seat and pedals. A wooden box occupied the space where the passenger's seat should be, the chassis and the blood red African earth beneath was clearly visible through those floor panel gaps and the deafening silencer-deprived engine could be seen vibrating wildly on its damper-less mounts through the space where the instrument panel should be. In short, it seemed I was to be transferred to the holiday camp in something resembling the skeletal remains of a mechanised dinosaur driven by someone looking like he'd been around to witness the demise of that long-extinct species.

With hair sprouting from every exposed inch including his sunken eye sockets, my chauffeur's only words during the ninety minutes it took to reach base camp was the grunt of the company name followed by another to indicate my place was on the box and that I'd be well-advised to hang on to anything not likely to give way during the trip.

The reason soon became obvious. Precious few stretches of the track had escaped being decimated by tropical rainstorms sweeping the area and when one of those unrutted stretches was encountered, my chauffeur would turn to me with a manic ragged-toothed grin, switch to cruise control, climb up through a hatch above his head and perch there leaving apelike bare feet to do the steering.

Only on encountering ruts the depth of shallow graves did he come down from his perch to grip the steering wheel with gorilla hands, ram a complaining gearbox into low ratio four-wheel drive and force our shuddering steed through a series of obstacle courses the average tractor driver would have thought twice about tackling.

After an hour of this jack-in-the-box behaviour, just to add a little spice to the proceedings in one of the rutted stretches he switched to cruise control, let the rut do the steering and leaped chimp-like from his seat to land on a sizeable python trying to make a break for it from another box placed right behind mine.

Having wrestled his captive back into the box, he then climbed back on the roof, pulled a pack of tobacco leaves from the pocket of what might once have been trousers and rolled one of the biggest, most pungent cigars I'd ever seen or been subjected to.

There he remained for the rest of the way, shrouded in a smoke screen so dense anyone looking on from afar might have mistaken us for the African Queen. Only on rounding a bend to sweep heroically into a compound of grubby bell tents surrounded by an impenetrable wall of thorn bush branches did he deign to come down to wrench the handbrake on, slew the vehicle to a skidding stop, leap from the driver's seat and race away into the surrounding forest through a gap in the thorny hedge.

Gazing drop-jawed after him as he disappeared into the trees, my reverie was interrupted by a voice sounding in my ear.

'You've met Miewkeus then.'

'Mucus?' I said, turning to address the chest of one of the tallest, thinnest men I'd ever seen.

'Yeah. Product of South African/Swiss parents born not far from here in one of Schweitzer's missionary stations, poor bastard. Never quite got over that,' he said in an accent that immediately betrayed his London East End roots. 'Mail?'

Chapter 3.2

Way ahead of you, I thought, handing him the package as he was speaking. No way I'm going to get caught out the way I'd been those first days in Ethiopia. Mislaying the mail was a very good way of finding yourself branded with some nickname you wouldn't want repeated in polite society.

The look on the thin man's face said it all. 'You've done this before, haven't you chum. Respect. Enough in fact for me to bother introducing myself.'

'Staff Serg… sorry… force of habit. Name's Jackson. Jacko for short. Quartermast… er… camp manager round here. Welcome to our humble abode. It ain't much but we call it home.'

Difficult to disagree with that. Took the words right out of my mouth had former British Army catering corps Staff Sergeant Jackson. Not much is the way I'd have put it too. The camp arrangement had all the charm of an army barracks, each army surplus tent planted in regulation ranks equidistant from all the others, every door pointing away from anything worth looking at.

Off to one side, a large tarpaulin-covered rectangle of poles cut from the surrounding forest had been implanted to act as a mess tent, a smaller rectangle added on to cover the kitchen facility. And the promised bar? A large refrigerator to one side of the mess tent containing nothing more than a small selection of bottled beer and the inevitable Fanta and Sprite, the oil world's concept of what constituted a soft drink.

The cuisine? Rather better, I hoped, than the absence of flesh on the man introducing me to 'home' implied. The Shell man in PG had 'promised' all the delights of a French kitchen. So where, then, were the gingham table cloths that covered the tables of every French restaurant I'd ever been in? In the wash?

The look on Jacko's face in answer to my own non-verbal inquiry suggested not. Gingham wasn't regulation army issue, it said. And neither, for that matter, were table cloths per se. Far too much trouble for a fighting force to bother with, a force that seemed quite content with bare trestle tables and easily transportable folding chairs.

Transportable? Was base camp going somewhere? Judging by the barricade of spiky vegetation surrounding it, it looked like this was its final destination. Anyway, why did we need a barricade in the first place? To stop the inmates going AWOL?

Indicating one of the regulation issue folding chairs ranged around the regulation issue mess tent trestle table, Jacko invited me to join him as he leafed through the sheaf of mail I'd brought. The fence, as he offhandedly put it, was not so much to stop us getting out as to prevent uninvited guests getting in.

Based on the concept of the East African boma, a ring of thorn bushes erected round rural villages to stop marauding big cats helping themselves to goats at night, the principle, said Jacko almost proudly, had been cunningly employed here in reverse. Here, the barrier had been erected to prevent local goats helping themselves to our food supplies, a deterrent that'd meant not having to post sentries at night. 'Any other questions?'

Yes. Where was the sea? All I could see from the camp was a stretch of decidedly dirty-looking khaki water that bore about as much similarity to the lagoon of picture postcards as this tented arrangement bore to the holiday camp image the man from Shell had planted in my head.

'That way,' said Jacko pointing distractedly to an indeterminate place beyond the 'lagoon'.

'Walking distance?'

'Not unless you don't mind getting there oven-ready. You'd need to make a serious detour round the mangrove to get to it. Not my idea of fun.'

Damn Mr Shell man. Just when I was beginning to change my mind about the oil companies, his description of facilities at the exploration crew's base camp was revealed in all its little oil company have a laugh at the contractor's expense joke glory.

So no change there then, I thought. Shell's attitude towards lesser humans drafted in to help with its endeavours was clearly no different to that of any other oil giant farming the dirty jobs out to contractor sub-species like me.

***

Settling down for a well-earned siesta in the tent I was to share with a colleague currently out on the lagoon doing a bit of echo sounding, I had time to reflect on what I'd got myself into. The only difference between this mob and the 3S outfit I'd 'served' with in Ethiopia was the range of nationalities. In Ethiopia it'd been mainly Yanks from the Southern States with a smattering of 'aliens' like me, Whackamole and the Viking. Here, it was the same but with Brits outnumbering the non-Brits.

I wasn't sure which was worse. At least in the company of Yanks I was seen as something of an eccentric and able to take advantage of that label whenever the need arose. Here, no such advantage existed. Step out of line with the milliterati who made up the bulk of the base camp personnel and you stayed out of line. Non ex-military were regarded with such deep suspicion no amount of eccentricity would save you. All it won you was a reputation for toffish superiority and a degree of fellow crew member indifference towards requests for cooperation to rival that of any goat intent on rifling the camp's stores.

So putting Jacko and his team straight on the language spoken locally probably wasn't the best way of making friends, influencing people and getting those with the power to make your life a misery work happily on one's behalf.

Gathering in the mess tent for dinner with the team that evening, once Jacko had introduced the new boy to the ten-member ensemble I settled down over the regulation issue cremated steak and chips fare to listen for an opening into what was inevitably heavily military-related banter.

From experience, regardless of what one had to contribute, I knew the absolutely worst way of integrating into the crew was through staying mum. All that got you was a reputation for standoffishness and with six full months of contract still ahead of me, that I could do without. To survive in the bush I needed the full cooperation of those in charge of the supply lines, the very people now seated round the mess tent table.

In the end I did get my chance to chip in, but not without the exact consequences I was trying to avoid. Having run through the range of reminiscences of life in the Royal Electrical and Mechanical Engineers – the army corps that was the primary reservoir for keeping OTC supplied with specialist personnel – the talk then turned to various aspects of the job in hand and, inevitably, to the shenanigans of the locals employed to do the survey donkey work.

With about twenty words of French – Gabon's official language – between them, those of the crew who'd been out on the water surveying the lagoon had had trouble making it clear to the local labour force that now was not the time to go ashore to chase the wild hog they'd spotted in the bush. There was only an hour of daylight left and the hydrographic team still had to get back to base some distance away.

'The bastards in my boat just went sulky and started jabbering in hottentot or whatever their lingo's called,' said Spanners, the highly imaginative nickname of the professional Yorkshireman doubling as the crew's chief mechanic. 'Couldn't understand a word and in the end I grabbed the tiller and navigated my own way back. Bloody bungs just sat there looking daggers.'

'Fon,' I said before I could stop myself and before he could enlarge on what was clearly headed for racial supremacist territory. 'Their lingo. Looked it up before I came here and it's not hottentot, actually. It's Fon.'

Shitshitshit, had I actually just used the word 'actually' in such company? If that didn't get me branded an aloof superior git and a target for sneering working class hero jibes, nothing would. The sort of aloof superior git who'd probably end up finding his bed booby-trapped for being a clever bastard.

In the end it didn't. But only because Spanners had seen a way of amusing the entourage by putting the smarmy newcomer in his place.

'Might be Fun to them,' he smirked to the general amusement of all at the table. 'When you've been here a bit longer, lad, you'll find it's just a pain in the arse.'

***

After that little exchange, suddenly the bush held a far greater attraction than anything base camp had to offer. Being removed from the salt of the earth that inhabited it – an appropriate moniker I thought for out and out irredeemable racists like Spanners who'd have done the world a favour by being left there – would most definitely be the more preferable option. While I might have to work a bit on getting them to fulfil requests for supplies at least I'd be spared having to bite my tongue every meal time. I could hardly wait.

Unfortunately I'd have to. And so would the Jungle Bunnies, the two fortunately non ex-military co-surveyors I'd been teamed with to undertake the forest survey work and who'd found themselves branded with a joint, almost inspired, nickname as a result of their identical initials – JB.

Before we'd be able to start the survey work proper, said

the Shell wheel who'd flown noisily and ostentatiously in by helicopter to brief us on the extent of work required, the first job was to open up an access line to the selected survey area.

Starting at a shoreline point at the far end of what was in fact a series of interlinked lagoons, the plan, he said, was to clear an old logging track by hand a good distance into the forest. For the initial stage we'd be commuting from base camp to the head of the track daily to oversee a debris-clearing team of locals and drive the cleared access track a bit further into the interior. Only when the commuting time across the lagoon and up to the workface began exceeding available clearing time would we be establishing a forward camp along the track and begin depending on supplies ferried to us from base camp. Once we'd reached the desired distance inland – about ninety kilometres – we'd then set to work carving our way into virgin jungle from a number of further forward camps to gather the requisite sub-forest canopy topographical data. Easy peasy, eh?

'On the face of it, yes,' said Jungle Bunny One, a rather serious Scot with more survey experience under his belt than JB Two and me combined, most of it in Nigeria's Niger river delta, 'providing Plotter's Law doesn't make an unwelcome entrance into the programme.'

'Plotter's Law?'

'Sure,' said JB Two, a behemoth of a New Zealander who'd joined the team for the sole purpose of earning the fare home. 'Ya never heard of it, the exploration surveyor's universal law of the jungle? A bit like Murphy's Law but with a better plot.'

'Sorry, not with you.'

'Better spell it out Two,' said JB One.

'Well, it ain't complicated,' said JB Two scratching the chin that lurked somewhere under an enormous bush of ginger whiskers. 'Plotter's Law states simply that what can fuck up will fuck up leaving the one being fucked up having to spend as much time unfucking up as actually fucking working. That's about it in a nut, er, Shell... Right One?'

'Pretty much,' said One through lips trying not to curl on Two's not especially well-veiled verbal poke in the eye of a superior who clearly didn't know his compass from his elbow.

'Simply put,' One managed at last, 'what Two is trying to say is that if Plotter stays away, supplies keep reaching us without interruption, we don't encounter any impassable obstacles, the natives and the wildlife remain friendly, no one's struck down with malaria or any other debilitating condition, no limbs are hacked off during the track clearing process and the labour force doesn't down tools on discovering the full extent of the work they're being asked to do, then we might, at a stretch, be able to get the job done by the end of the decade. But even that might be pushing it,' said One with a rare, wry, I'm only half-joking look in the direction of a man who could probably have him flogged for insubordination if he felt like it.

'You have three months,' said the man without smiling. 'It's now July. We need to have enough topographic data available to start planning the full seismic exploration phase in the final months of the year. So I'd suggest you get your arses in gear tout suite and start chopping. Capiche?'

JB One looked at JB Two, JB Two looked at me and we collectively looked at The Wheel trying not to laugh.

From past experience of survey jobs of one thing we were jointly and severally certain. Fail to include Plotter's Law in the work schedule equation and the one doing the schedule arranging would, as every surveyor worth his salt knew, be fucked. Good and proper. If he escaped developing a serious heart condition on seeing the job totter from one calamity to the next leaving his lovingly manicured timeline spreadsheet good only for lighting fires, it'd be nothing short of a miracle.

Should we bother pointing this out to the neatly booted and suited company man with the clipboard, the trimmed moustache and the helicopter?

Nah, the looks that passed between us said. Time taken doing so would probably be about as well-spent as bothering to argue the toss with the camp racists. So no. He could find out in his own sweet time. It'd be good for his education. Anyway, we had better things to do.

***

Things like writing our wills. While our employer and their client might be trying to pretend all would go swimmingly out in the forest, having seen what we were faced with, the two Jungle Bunnies and me weren't so sanguine. The area we'd be hacking our way into had 'Here Be Dragons' written all over it and in this case that may very well not be just a metaphor. There were any number of known health and welfare-threatening demons out there just waiting to wreck our best laid plans and even more unknowns. So, in the light of the obvious absence of back-up for the poor sods being sent into the lions den with no means of defending themselves, the writing of a will seemed not just prudent but perspicacious.

Not that any of us had much to leave anybody. We were, after all, just bum contractors with hardly a pot to piss in despite existing on expenses and company-supplied vittles for most of our time in the field. What little salary we got as mere supernumeraries to the moneybags oil industry had, it's true, gone straight into the bank untouched. But by God we drew down on it in the weeks, sometimes months, between jobs. It wasn't just the rent and basic living expenses that had to be covered. There was also the small matter of one's in-between-jobs bar bill, a not insignificant amount in the case of those with an oil industry involvement-created drink affliction – i.e. about ninety percent of the oil world's workforce.

Nevertheless, it did seem a good idea to jot down the names of a few beneficiaries to be the recipients of our worthless assets, as much for the comfort factor as anything. Being a beneficiary, those named would at least feel obligated to remember us in some small way should the forest swallow us up without a trace.

I do not use the word lightly for swallow is exactly what could be our fate in territory bursting at the seams with swamps, crocs and pythons big enough to ingest the ego of even the highest level oil company executive.

And that was without even mentioning Legba, the decidedly grumpy Voodoo spirit of the forest local folklore warned could easily take offence over having its equilibrium disturbed. Put Legba's nose out of joint and we could quite easily find ourselves joining those who'd gone before. The ones who, as we were eventually to discover, had set foot in this forest some years before and, if our local labour force was to be believed, had never been seen or heard from again.

Chapter 3.3

The existence of the hole the Landrover was now at the bottom of rather confirmed we weren't the first ones into this forest.

The loggers we already knew about. It was their track, now thoroughly overgrown and almost disappeared in places, we were opening up. But the area the clearly man-made Landrover-sized hole was hidden in was different. Out of the dense dark forest we'd emerged into the blinding light of a long wide treeless plain that could only have been carved out of the jungle for one reason. This had once been an airstrip.

But created by and for whom? We'd encountered nothing like it in six weeks of battling our way up the track, detouring round prostrate trees wider than a man was tall, building tree trunk bridges across teeming blackwater torrents and setting up three staging post camps along the track. So stumbling on it during a lone recce ahead while the other two Jungle Bunnies were otherwise engaged had me scratching my head and punching the air in delight at the same time. For the fourth camp we'd be able to see the sky for the first time in weeks and I almost jumped for joy in the driver's seat.

Then my world turned upside down and I was left jumping for an altogether different reason – to escape the attentions of my slithering uninvited guest.

I'd been wondering when the first major setback would hit us. So far, after commuting from base camp across the lagoon and up the track for the first two weeks then establishing three forward camps over the next four nothing especially untoward had befallen us. Well, nothing if you discounted being eaten alive by mosquitoes and tsetse flies. For all we knew we could be being infected with any number of horrible tropical diseases despite the company-provided anti-infection drugs, one of which immediately became suspect on its sudden removal from the base camp medicine chest without notice or explanation.

We'd been told it was to combat two locally prevalent diseases – the mosquito-born filaria which could result in parasitic worms eating their way through your optic nerve and the debilitating tsetse fly bite-caused sleeping sickness.

Now the stock of unidentified little yellow pills had gone mysteriously missing, all crew talk revolved around whether its absence made us less safe or more. While being unlikely to suffer the unknown side-effects of a drug everyone suspected was being tested on us, not having it left us fully vulnerable to the worms the man Miewkeus said would eventually erupt through your skin and eyes if left untreated, and to bouts of falling involuntarily asleep on the job, never to wake up if you were very unlucky.

Had that just happened to me I wondered after scrambling dazed from the Landrover via the cab's sliding rear window? I'd accumulated enough tsetse bit scars to warrant concern that I might drop off at the wheel at any time. Is that why I hadn't spotted the hole the Landrover was now in?

Gazing round trying to get my bearings and figure out whether it was me or the shoulder-high grass that was to blame for our predicament I was about to decide on the latter when another worry superseded the first. Advancing on me through the vegetation with crazed looks in their eyes were the five Gabonese labourers who'd been catapulted from the back of the pickup into the grass around the hole.

Brace for a violent attack was my first reaction, retribution for an act of driving negligence that'd turned them into involuntary human Exocets. Then I saw looks of relief replace what I'd taken for menace and relaxed. I wasn't the only one knowing that had I suffered serious injury from the crash there was only one place the blame would come to rest. With giants like Shell well known for looking after their own staff first, their contractors second and the local labour force engaged by their contractors a very poor third, they'd be counting themselves extremely lucky if they escaped lengthy periods of incarceration for a/ failing to protect their immediate employer from coming a serious cropper in an environment they knew a whole lot better than he did, and b/ being the easiest target on which to pin blame.

So after being assured I'd come out of it more or less intact, their collective response was to first breath audible sighs of relief and then to burst into huge grins of unbridled joy. Without the Landrover, now lying sorrowfully half in half out of the hole, of one thing they were sure. They'd be getting a few unexpected days off, days in which they could go back to their villages to be treated like returning heroes coming in out of the bush with a story they could dine out on for years.

But not just yet. As every one of them knew full well, it'd be more a case of a premature end to their employment unless they applied local knowledge to helping their employer out of his current predicament.

His immediate need was for a rescue service of some sort. That they knew without being told. So as I surveyed the damage and the scale of my plight, they convened a committee meeting.

Breaking from their huddle after a few minutes of animated discussion in the local language, they first looked at me, then jabbered a few unintelligible words in my direction and collectively disappeared into the forest leaving me trying to work out what message had just been conveyed.

Going for help, I sincerely hoped, not a communal decision that the best course of action in the current circumstances was to make themselves scarce until the heat died down.

Either way, I now had no option. Unable to do much about my present situation alone there was nothing else for it but to resort to the exploration surveyor's regulation response to calamitous mishap and impending doom. Erect a table and chair and make some tea.

Sitting sipping at it while gazing out towards a herd of buffalo grazing at the far end of the plain, I began to speculate on what sort of rescue service my gang had hopefully gone in search of, all the time trying not to allow the possibility of having been abandoned in the bush to enter the equation. I didn't much fancy having to fend for myself for days on end until the Jungle Bunnies noted my absence and dispatched a search party to look for me. Forest buffalo were well known for taking great exception to having their territory invaded.

Top of my rescue service list was the drafting in of a band of locals armed with a rope. What else was there in these parts? A tame elephant prevailed upon to help, perhaps? Unlikely. African elephants, especially forest elephants, were notoriously difficult to domesticate so that option was quickly discounted. Ditto drafting in the equally notoriously bad-tempered buffalo.

The only other option could be the silverback gorillas spotted in the bush on occasion. Could they be tempted by a banana or two to lend a hand? Again, probably not. Being a target for local hunters their cooperation seemed unlikely and as the range of options diminished I began to wonder how we'd ever get my steed out of the hole.

***

With something I'd never even thought of, as it turned out – another beast I heard long before I saw it crashing through the trees and out on to the plain.

I'd heard there was another crew similar to ours trying to map a far swampier area some distance away but had never encountered them. Well now I was about to… and one of the mechanical monsters they rode.

Dropping everything the moment he heard one of his kind was in trouble a chirpy geordie called, of course, Geordie had diverted straight to me, booting his swamp buggy colossus to its maximum speed of 4mph to see what could be done.

Within moments of thundering into sight he'd hitched a line to the Landrover, whipped it out of the hole and over tea had started filling me in on how, after seeing a group of locals gesticulating wildly at him, he'd been alerted to the problem. With the locals unfamiliar with the geordie dialect it'd been down to a lot of charade-playing and arm waving to get the message across and with one of them on board to point the way, Geordie had got to me in ninety minutes flat.

'Bloody glad you did,' I said. 'Didn't much fancy spending the night here with those grumpy boeffs on one side and an angry mamba on the other. Not especially conducive to sleep.

'What really amazes me though is how the message got to you so quickly. It's only been a few hours since the crash and you were miles away.'

'Bush telegraph,' said Geordie. 'You didn't hear the drums?'
'No.'
'Nah. Me neither.'

***

Whatever the answer was we, we knew, would be the last to hear about it. It wasn't the first time something inexplicable had happened in this forest and asking for an explanation via the one or two labourers on our respective crews with a little French had never elicited much in the way of an intelligible response.

So by and large we left it, on this occasion having to be content with the tap-of-the-nose grins from every one of the crowd of locals who'd materialised out of nowhere to watch the recovery process, every one of them with a different suggestion on how to go about extracting the forlorn Landrover from the grave that seemed to have been dug especially for it.

Forlorn was definitely the word for it. A poor battered and bruised vehicle with a front wheel drive shaft now sticking incongruously up through the cab floor, a grotesquely twisted front bumper hanging on by a thread and two shattered headlights.

Would it still drive? Miraculously, yes, providing you didn't expect any help from the front wheels and night didn't descend before reaching the planned destination, Geordie's camp a few miles off the track we'd opened up. From there, said Geordie, we'd be able to radio Spanners with a plea to come and fix what, without four-wheel drive, was a near useless machine in jungle conditions.

Just how useless I was to find out the hard way en route to my rescuer's base. Having to make it across several muddy rivers with just rear wheel drive was a test of both my driving skills and the power of prayer but somehow I made it, the warped un-disengagable front wheel drive shaft gyrating wildly up through the smashed floor panels the entire way and threatening to gouge a hole in my leg whenever there was a need to change gear.

And my cutting crew? Let joy be unconfined, the looks on their faces said when the Landrover popped out of the hole and I informed them they'd have to look after themselves until I could get it fixed and retrace my footsteps to them.

I didn't need a translator to understand their delight on being left behind. With a sizeable village not far away, it'd be the fatted bush meat and the choice of unmarried girls for them for the next few days washed down with the crate of beer that'd somehow gone missing from the Landrover in the course of the kerfuffle.

At the bottom of the Landrover grave, said Jean-Baptiste, the gang's cutter-in-chief. *'Vous voulez l'a retrouver?'*

Not bloody likely, as Jean-Baptiste knew full well.

***

Ever since discovering beer in the fridge of the 3S crew in Ethiopia and seeing what it did to those with time on their hands in such remote places, I'd been wondering why alcohol of any description was allowed on any oil industry site. If HJ was to be believed it certainly wasn't on his rig in Libya, but that might've been down to alcohol not being officially anywhere in the kind of Islamic state to which my erstwhile friend had been posted.

'Recruitment bait,' was base camp manager Jacko's reasoning. 'There's precious few willing to give up the comforts of home for this life as it is. Take their beer away from them and there'd be even less. That's why in this neck of the woods we get a booze ration from Shell. A bribe to get workers to sign on for a posting to what everyone knows has a reputation for being the White Man's Grave. I reckon Shell's done the sums and found the pros of keeping the fridge full of beer more than outweigh the cons.'

No one on the crew was complaining... except the little sub-crew to which I was attached. To us out in the bundu it was a constant source of irritation. Not because there WAS a beer ration but because in our case, more often than not there wasn't.

It wasn't that we weren't allowed to drink while bivouacing out in the forest. We did when the occasional crate was included in our supplies. It had more to do with there frequently being no such crate, an issue not disconnected to the ingrained divergence in social nicety existing between us and those tasked with keeping our supplies coming.

In short, it all came down to background. On our crew there were those with a military past and those without. We three out in the jungle were firmly in the latter group, the one the former group regarded with extreme suspicion for bringing what they saw as hippie tendencies to their well-ordered little community.

In their eyes we were, quite obviously, fifth columnists on a mission to infiltrate and undermine their cosy little closed-rank society and as such failed to merit being given so much as the time of day beyond the basic information of it being one of either day or night.

So, with those parameters set it came as little surprise to any of us when calls to base camp from our antiquated army radio to beg for supplies would not infrequently go something like:

'You want what?'

'Food and water.'

'Why?'

Brief pause as survey camp caller rolls eyes at microphone in jaw-flapping exasperation.

'So we can use the food as fish bait and the water to drown the fish in (dickhead – mouthed silently at the mike).'

'I'll have to get clearance for that.'

'Well while you're getting it, could you also ask if it'd be OK if we supplement the food stock with one of the labourers? Got one here who's becoming a pain and since the cupboard's now bare thought we might kill two birds with one stone.'

'Negative. You're not allowed to kill birds.'

'Damn. Ah well, it'll have to be the panda then… since we've got NO BLOODY FOOD!'

'Roger that. Stand by ten.'

Ten minutes to the second later in which the radio battery drained almost dry it wasn't unusual for the base camp radio man to return with the news that a re-supply was being organised and would be with us on the next scheduled shipment, ETA three days time.

'Meanwhile, the quartermaster suggests you spare the birds, the panda and the labour force and try the local food. Out.'

With the local food comprising anything our cutting crew could catch in the forest and this level of urgency being base camp's invariable response to our requests, such calls left our little gang plotting ever-darker deeds against those who consistently assured us the needs of those putting their lives on the line to gather the required map-making data came top of their priority list.

So, if giving us priority was truly the case, we asked ourselves, how come, while they were tucking into the regulation base camp steak and chips washed down with fridge-cooled tall frosty ones, we were left to survive on snake, monkey, bush pig and muddy, bilharzia-ridden river water? A character-building exercise, perhaps? Did they think the hippies out in the bush needed toughening up? Or was it that surely, as hippies, they'd be used to living off the fat of the land?

As we moved out of the track clearing phase of the job and into the data gathering, bush-bashing phase, the jury was still out on that.

Chapter 3.4

Given the level of priority base camp accorded to keeping those out in the field well-supplied, Spanner's response to our request for him to come and fix the Jungle Bunny Landrover was nothing if not predictable.

'Fuck off', was his initial reaction to the call. 'Come all that way just to fix a drive shaft? You can get it back to the lagoon, can't you? We can fix it there.'

Under normal conditions he'd have had a point. Getting the Landrover to a more accessible place would definitely have been possible in, say, the leafy lanes of Spanners' native Yorkshire. But here, as I tried to point out as diplomatically as I could to someone who only left base camp to go on leave or on fishing trips in the lagoon, in case he hadn't realised it these weren't normal conditions. We weren't in Yorkshire's leafy lanes. We were in the darkest heart of darkest bloody Africa and any number of mishap-making bear traps lay between our current position and the lagoon. I knew, I'd been instrumental in constructing ways of getting round them... but only in a fully functioning vehicle with four-wheel drive and high and low ratio gearing.

So if he'd be so kind, since our chances of being able to get the vehicle to the lagoon were in the vicinity of zero, perhaps on this occasion he'd be good enough to get off his fat lazy racist arse and bring his tools and skills to us in his own, barely bush-tested, Landrover.

In the end he grudgingly agreed, but only after being reminded of something that had the effect of focussing his mind nicely. Unless he made the trip to fix a vehicle on which we in the field were completely dependent, I told him, it wouldn't be just us not making it home for Christmas. Since every member of the OTC crew was contracted to remain on site until every one of Shell's required survey tasks was in the bag, if we missed the pre-Christmas deadline for completing the sub-forest canopy data-collecting part of the survey everyone would be stuck out here with us and that'd mean just one thing. He'd likely be getting home to an empty house, his wife having fulfilled her threat of leaving him if he missed one more Christmas.

With Spanners going very quiet on the radio, one word sprang to mind. Checkmate.

I was right. For within hours of receiving the call, our now fully compliant chief mechanic had organised a rescue mission, piloted the base camp vehicle transport barge across the lagoon at breakneck speed and all but raced up the track to reach us almost before we'd finished dinner.

Isn't it amazing, we Jungle Bunnies commented to one another as he set frantically to to get our vehicle back in full working order in four hours flat, how the prospect of missing Santa can energise some folk. Maybe we should apply the same threat to the food re-supply problem. Not just for our benefit but for that of the local wildlife, the numbers of which must surely have suffered as a result of our unprovisioned presence.

***

Actually, the forest told us as the foot-slogging part of the data-collecting phase of the job got underway, your presence here has had less of an effect on wildlife numbers than you might think and went on to prove it with every foray we made into the bush. Hardly a day went by without one of us having to take refuge up a tree to escape sharp-tusked wild boars rampaging blindly through the undergrowth, narrowly missing treading on fully-formed vipers as one's attention was distracted by mad hornbills and monkeys crashing around in the treetops, or coming face-to-face with sizeable snarling civet cats guarding their lairs.

All these and more were such everyday occurrences that in the end they barely merited a mention to the others on returning from sorties into the forest from the four forward camps we'd established. Most exchanges related to the more headline-making news of obstacles stumbled upon and the misadventures befalling us on trying to find ways around them.

Ranging from having to make wide detours around families of forest elephants jealously protecting their offspring to being halted by areas of swamp into which no one with any sense would ever venture, such obstacles had every one of us cursing them almost as much as the people behind a survey plan that took zero account of the possibility of encountering impassable sections.

The plan, if that's what it could be called, was to head undeviatingly into the forest along a compass bearing for some fifteen kilometres recording ground conditions and elevations as we went. Distances were to be measured with something called a Topofil, little more than a child's toy comprising a spool of thread and a basic odometer in a plastic box, elevations were to be recorded using a simple clinometer and to stay on a compass bearing we had, well, an army compass. Once started on a bearing, went the plan, under no circumstances were we to deviate from that bearing, a requirement that left us no option but to circle round obstacles and try to pick up the line on the far side, no matter how extensive that obstacle was.

It took every one of us just one bush-bashing foray to discover the extent of lunacy in this order. All too often the obstacles encountered were so extensive it made picking up the line on the far side a matter of guesswork and even, on occasion, to having no option but to estimate the likely lie of the land beyond the point at which no further progress was safely possible.

Such was the case on one line I was surveying that led to a detour so wide it left me so lost and disorientated in the forest I began to think that wherever this was could well turn out to be my final resting place. Of only one thing was I certain once the thought struck. It was down to me and me alone to find my own way back to known territory. With the line's start point a virtual unknown thanks to having only the Landrover's highly inaccurate odometer to tell us where along the cleared logging track this line had been begun, the chances of any search party finding me was next to zero and that also went for my cutting crew, every one of whom had had a different idea of how to get round the massive area of swamp we'd hit about a kilometre short of the finish line.

With every one of them heading off in a different direction I was left having to choose which to put my faith in. In the end it was Albert, a youngish lad promoted from basic bush basher to Topofil technician on the grounds that none of the others seemed to have the sense to tie the thread off at regular intervals to avoid losing the line if the thread broke.

Two hours later with Albert obviously as disorientated as me, I rapidly reappraised the decision to have put so much trust in his forest navigation skills and took the executive decision to stop to regroup our senses over a spot of lunch, in his case a stick of manioc – a cassava-based bush snack – in mine a can of sardines.

Then the first disaster. No key to open the can with.

No problem. I'd use my knife, a lovely bone-handled blade of forged Sheffield steel that'd hung faithfully on my belt the length and breadth of Africa just waiting for moments like this to return the favour of being rescued from a Brighton junk shop.

Oh dear God. No knife. Must have become detached somewhere between here and our third bivouac camp along the trail about three clicks back, the last time I'd used it.

Feeling frantically round my belt in a failed attempt to locate it I felt panic rising. Being lost in the bush was one thing. Being lost in the bush without a knife – THE premier piece of any exploration surveyor's kit – was disaster incarnate. As JB One had once remarked on inspecting my pride and joy, 'a surveyor without a knife is like a man without a dick' and now I knew exactly what he meant. I felt both desolate and emasculated without it.

Fortunately both I and Albert still had our machetes, mine with a point still sharp enough to hack my way into the sardine tin. But that didn't solve our immediate problem. We were still hopelessly lost and Albert wasn't proving much help, the sweat on his furrowed brow suggesting he was as close to panic as me.

'Oh good,' I thought as I downed my lunch with grubby shaking fingers. 'Now I've got a frantic passenger to look after as well as myself.'

***

In the end we did extract ourselves from the predicament but it was more by luck than judgement.

Priority one in such a position, I knew, was NOT to panic. What this needed was a clear head to be able to think. So after first checking the ground carefully to make sure my backside wouldn't be disturbing any well-camouflaged biting beast I sat down under a tree, closed my eyes and cast around in my mind for any advice I might once have been given for dealing with situations like this.

'Eat your vegetables!'

Eh? What was my mother doing here?

'Helps you see in the dark.'

Well, yes, that would be a boon in this light-deprived place and I had mentioned the dearth of fresh vegetables in my diet out here in a letter to a friend back home, receiving a packet of radish seeds in return. But that didn't solve my immediate problem.

Or it didn't until I thought about it. Mother's vegetable advice had clearly been my mind's roundabout way of prompting recall of something that friend had once said about problem solving. A man of unconventional otherworldly philosophy and lifestyle, he'd once told me that, to him, the best way of solving a seemingly intractable problem was not to think about it. To clear the mind and let instinct and intuition do the talking.

Had my friend been expressing that sentiment a few years later, to get the point across he might well have summarised it by quoting the 'feel the force' direction issued to Luke Skywalker in the first Star Wars movie. But at the time, since we didn't yet know about Obe-Wan Kenobe and his Jedi Knights, the message had had to be conveyed in more formal language. Language I now found resonating in my head.

Should I apply the process to my current situation? Would it prove more effective than staying faithful to the proven science of compass-based navigation? As one with a science background I knew I should be choosing the latter option. But since it was that option that had landed us in our current predicament in the first place, what real choice did I have?

The more I dwelled on it, the more it seemed the only way to go and in the end I opened my eyes, put the compass in my hand away and gave the one in my nose a chance.

'This way,' I told Albert with as much authority and confidence as I could muster, plunging headlong into the undergrowth and hoping to hell the break my nose had once suffered on the rugby field wouldn't affect its direction-finding capability.

It didn't. In twenty-five minutes flat we were back on the trail we'd been cutting, and it was all thanks to that stupid Topofil toy I was now hugging in delirious gratitude for getting me out of what could have been a very serious pickle. Had it not been for the Topofil thread I found wrapped around my leg, we could well have walked straight across this lightly bush-bashed section of the trail without noticing and ended up going round in circles forever.

Untangling myself from the line, as relieved as I was to have 'found' myself one question still nagged. How the hell had we set off in one direction, diligently following a compass bearing that should have led us round in a semi-circle to the other side of the swamp yet ended up a hundred metres back from the point from which we'd started? It made absolutely no sense, and no matter which direction I attacked the conundrum from on the way back up the trail to Jungle Bunny central – having taken the executive decision to abandon this line a kilometre short and simply insert the words 'Here Be Dragons' after the survey readings up to the point at which we'd hit the swamp – I was still trying to figure it out as we dragged ourselves hugely relieved back into camp… to be greeted by the rest of a now equally relieved (and shamefaced) cutting crew, every one of which had gone through roughly the same experience as me and Albert.

***

The 'Here Be Dragons' characterisation of the forest's swampier parts wasn't as far off an accurate description of the place as it sounded… as we were to discover on our next, and final, sortie into the bush a week later.

First it was the size of the monitor lizards – seen to increase the closer they lived to water saturated areas – that told us we were approaching a low-lying soggy quagmire section of this line. Then there was the little matter of the increasing prevalence of one of the lizards' closest living relatives… the undisputed king of misnomers, the rock python.

As the largest of the python family growing to several metres in length, it was not only known to prefer fetid marshy conditions to rock formations but had markings unlike any rock I'd ever seen bar serpentine, a metamorphic rock named, naturally enough, after its red/green snaky colouring.

Had there been any serpentine in this neck of the woods, the rock python might well have used it to camouflage its presence. But since there wasn't, it usually chose the brown/green vegetation near swamps in which to lurk, awaiting something edible to pass by. Something like us.

By and large rock pythons usually left adult humans alone, the average-sized python having the sense to realise this prey was too big for even it to get its head/mouth around. But with virtually everything defying the law of averages in this forest there was always a better than average chance of encountering rather bigger than average members of this species and on occasion we'd catch fleeting glimpses of such over-sized beasts sliding out of the way of our direction of travel.

With none of us ever seeing more than a tail disappearing into the undergrowth a guarded sense of security had developed amongst our team, the racket we made cutting a path through the forest clearly working to scare any lurking python off.

We reassessed that belief the day on that sortie that one of my gang found his leg in the mantrap-sized mouth of a veritable monster, one so big it knew it was all but invulnerable to anything this forest could throw at it and as such had no intention of vacating its nice moist hidyhole on the fringes of the swamp our line of oncoming traffic was approaching.

Big mistake on the python's behalf. It'd probably never come across traffic bearing razor sharp machetes before, especially traffic with such a lust for blood the paths we were cutting would have resembled the aftermath of the Trojan wars had those in our way been stupid enough to hang around to see what was coming.

As the captured gang member shrieked and struggled to escape becoming the monster's lunch the others set upon it like hyenas on a downed wildebeest, hacking and battering the beast until it released its hold and tried instead to bite back.

Another big mistake. All that did was incense its attackers, blades flashing until it succumbed and finally breathed its last bloody gasp.

After giving it a moment to be sure it wasn't just lying doggo, the gang then unfurled it to gauge the extent of their triumph, issuing little gasps of their own as they did so. At well over eight metres long, this was the biggest of its kind any had ever seen, my cutter-in-chief Jean-Baptiste told me. Every one of them was a forest local and despite seeing many pythons in their time, none had ever come across such an immense specimen.

That didn't mean they didn't exist, J-B added. It just meant that no one who'd had the misfortune to encounter one had ever lived to tell the tale... or been seen again.

How very comforting, I thought with a shudder. So all this time, anyone straying from the group – as I had just a week earlier – could easily have found himself falling victim to such a beast, the prospect of which led me to instituting an immediate change in my working practise.

The cutting gang wasn't surprised to find their gang boss sticking closer to them now than at any time during the past months, eyes fixed firmly on the undergrowth. His already white face had paled even further after the monster python encounter, clearly on the realisation that he was, after all, mortal. More mortal than he'd felt at any point on this mission or, for that matter, at any time in his past twenty-five years.

This was real, his ashen face told the gang. This was happening. It was no longer something he could feel a sort of TV viewer detachment from no matter how many mosquitoes bit him. Put one foot wrong and he now knew with absolute certainty it'd be him, not some film actor, suffering the consequences of not keeping his mind on the here and now.

For which read not keeping his eyes on the undergrowth. After what he'd just witnessed he'd clearly decided that no tree-inhabiting danger could compete with what might be lurking at their base. The imagined horror of finding himself in the coils a beast like the one he himself might have stumbled upon left no room for diverting any attention away from the tangled mess at his feet.

With every sense now trained on it, the picture now in his mind's eye was clearly one of finding himself with time to envision the fate that awaited as those coils tightened slowly around him.

Was that a small smile spreading across Jean-Baptiste's face as he read what was in his boss's mind's eye? Undoubtedly. And I didn't need a translator to interpret the words he was now mouthing to his fellow workers.

'Who's the boss now then?' was his unmistakable jibe at my expense. 'What's the betting that once this sortie's over that's the last we'll be seeing of him in this neck of the woods? Any takers?'

Chapter 3.5

Whether any of the gang took Jean-Baptiste's bet I have no idea. But since I wouldn't have bet against it myself, it seemed unlikely. The man's reading of my reaction to the events of the past few days was as close to being spot on as made no difference... although I'd never have admitted it to myself or anyone else at the time.

Nah, 'course my nerve wasn't shot was the image that had to be portrayed to my fellow OTC crew members. Like them I was a full-on, red-bloodied, testosterone-fuelled, rufty-tufty oilman. The sort that didn't have nerves. Just wiring. And anyone saying any different might just like to step outside to have it proved to them. Grrrr...

In truth it wouldn't have been much of a fight. Deep down I knew that nerve had definitely received a bullet. The question was, was it a cannon shell or just buckshot? The latter and there was still the chance of a full recovery. The former and there'd be no escaping the fact that the shelf-life date everyone in this industry had stamped on them had, in my case, been brought forward somewhat.

I wouldn't have long to wait to find out which. There was another swamp up ahead that needed skirting before we could press on to cover the remaining ten kilometres of this line.

As things turned out, whatever was left of my nerve wasn't to be tested to anything like the extent expected. Well, not immediately. With the swamp only about fifty metres wide we traversed it in minutes to find the forest on its far side way less dense than it looked. Progress through it was swift and got swifter as the tree cover continued thinning.

'This can't go on,' I recall thinking at the time. 'Legba's just playing with us. Pretty soon, once the spirit of the forest sees we've become complacent, he'll make something awful happen.' I'd been down this road before and knew for a fact that now was the time to hone the senses, not relax them.

I was right. With every sense now at maximum torque I began picking up a noise I'd never heard before in this forest. It was like huge breakers crashing on some distant shore and the moment I sensed it I froze. Something very very big was ahead of us and I looked beseechingly to the gang for advice as to what our reaction should be.

Gone. Every one of them. Sprinting for their lives through the now just woodland dense trees TOWARDS the thing.

'Ah well, in for a penny...' I said to myself breaking into a half jog to catch them up. 'They're the forest experts. If they think this is the thing to do, who am I to argue?'

On they ran, dodging between trees and starting to accelerate with what I took for whooping war cries until they were all but out of sight. Fortunately, with my trusty Topofil carrier Albert amongst them, to avoid getting lost again all I had to do was follow the thread.

Then I saw it, the Topofil box left lying on the ground, abandoned.

'Ye gods,' I thought as the whoops intensified ahead of me. 'Whatever it is, they've caught it!'

Right again although I couldn't immediately see exactly what, the dazzling sunlight blinding me momentarily until my eyes adjusted from all those months in the forest gloom.

When my vision returned and I saw what 'it' was I froze again. This was some kind of collective mirage, surely. A little Legba joke designed to make us think all our travails were over when, in fact, they were really just beginning.

My gang had caught it all right. The Atlantic bloody Ocean. And without a thought for any lurking riptide current or other submarine menace, they were now dashing gleefully across the wide, pristine white sand beach to dive headlong into the pounding surf and frolic joyously in the shallows shrieking like over-excited schoolgirls.

***

Setting up camp on the beach and preparing to feast on the two days of rations we each had left after completing what I'd estimated would be a three-day outward sortie in a day and a half, I couldn't help but wonder what the crew of the oil tanker cruising offshore had made of the spectacle. Blasting its horns as we danced around in the surf waving maniacally at it, it was clear someone had had binoculars on the band of ragged lunatics who'd burst out of the forest and raced lemming-like towards the breakers to hurl themselves fully-clothed into the briny.

Fire ants might have been their first guess. Find yourself wandering blindly into an army of those and a close-by ocean could save your life. Then again, they could be escaping some hostile force chasing them through the forest. Should they heave to and launch a rescue mission? Nah. By the time they'd brought the football pitch length tanker to a halt they'd be miles away along the coast so maybe best to just alert the authorities ashore and leave it to them.

If that's the decision the captain on the bridge was coming to, it occurred to me that any rescue mission launched wouldn't have much trouble finding us. All they'd need do was follow the smell. Months in the fetid pungency of the forest had left its mark on us – as had frequently been alluded to during R&R breaks at base camp and Port Gentil – leaving all those subjected to the Jungle Bunny presence scrabbling around for excuses to be somewhere else.

Well, pretty soon they'd be having to dust those excuses off again, for as soon as we got back from this sortie we'd be packing up and making our way down the trail to the lagoon for the very last time. Every required line had now been cut and 'surveyed' and it was time to deliver the final package of data back to our masters and take a well-earned break. Not least to find a bath to soak in to rid ourselves of both the forest stink and the animalistic tendencies the absence of civilised living facilities had reduced us to.

As we all but jogged the twelve kilometres back up the already cut trail to Jungle Bunny central I had time to revel in the thought of that upcoming bath and to reflect on just how far into animalistic territory we'd sunk.

A mighty long way, it occurred to me as we passed the lifeless coils of that monster python now all but stripped down to its skeleton by a local colony of army ants. At the mention of 'coils' to myself, a wholly unconnected event in which coils also featured had materialised from nowhere. An event of such unbridled animalism, it summed up just how divorced from the so-called 'civilised' world we'd become.

Returning to camp beaming with relief from one of his more successful visits to the bush to answer an urgent call of nature, the Kiwi colossus that was Jungle Bunny Two hadn't been able to contain himself.

'This you GOTTA see, blue!' he proclaimed through his monstrous ginger beard, grabbing me by the arm and dragging me protesting to the scene of his 'crime'.

'Get an eyeful of THAT, mate! Gotta be some kind of record, eh?'

He wasn't wrong. The pythonesque coil he'd left on the ground put anything I'd ever produced to shame, even the ones disgorged on that airstrip in Ethiopia as the supply plane was coming in to land or here while, in full squat-position defecatory flow, all I could do was watch and sweat as a fully grown scorpion advanced purposefully through the leaf litter towards my unprotected genitalia.

Damn right, I agreed, somewhat in awe. Definitely some sort of record.

'You betcha! Reckon that needs recording for posterity, mate. Got any film left in your camera?'

I had but was highly reluctant to use it on still-life studies of this nature and told him so. With my precious film stocks running perilously low and unable to get any more until our next leave in PG, what I had was being reserved for recording this mission's more action-packed Jungle Jim moments.

'Ya bloody meannie. C'mon mate. Just one measly shot. That's all I want. I'll pay yer.'

Me a meannie? How dare he? Which one of us here had never bought a round in all the months we'd been in Gabon? Which one of us spent all his leave breaks in his hotel room living on rations smuggled to town from base camp so he wouldn't have to fork out on restaurant food? Which one of us had spun the line that he had no option but to conserve his money on account of his being so averse to economy class air travel that every time he flew he had to shell out on upgrading to first class? Not me.

So pay me? Don't make me laugh. Getting paid for the picture was about as likely as me believing any airline would let a man-gorilla dressed in raggedy t-shirt and shorts with home-made sandals fashioned from a tractor tyre on his feet and a pillow case of his miserly possessions slung over his shoulder anywhere near the plane, let alone take a seat in first class.

'Oh come ON, blue. Thought we were mates. You know I'd do it for you if I had a camera.'

Oh for God's sake. Knowing he'd never stop whining about it for the rest of the job if I continued to resist, in the end I just pointed the camera and snapped. Anything for a quiet life.

Not so quiet when it came to getting the film developed in PG. Opening the envelope containing the prints, first my jaw dropped then JB Two's. Nothing. Every frame was ruined. How? Why?

With every other roll of film developed and printed OK, it had to be down to the dunking my camera had got on slipping off the log bridge spanning a wide stream within spitting distance of getting back to the final Jungle Bunny camp of this mission and going tits up in the almost opaque water. Despite being submerged for no more than a couple of seconds, the water must have got in and ruined the film.

It wasn't the only thing it ruined. The camera resolutely refused to focus properly after that, a deficiency that meant that was that for recording the rest of this job on film.

'Ha! That'll teach you to take your mind off the job in my forest,' I heard Legba laughing in the back of my head. 'Thought you were home and dry, did you? Thought you could allow yourself the luxury of letting your mind wander on reaching that log bridge, did you?

'Not in my forest, my friend. Really bad idea to take old Legba for granted. Bad things happen if you do.'

***

It was a lesson well-learned. Although I had a sheaf of photos documenting almost everything encountered in the forest up to the camera dunking disaster, I'd have nothing with which to record the final phase of this survey job, the one likely to be the most photogenic. Before we packed up and left the country Shell wanted us to survey the region's primary river, mapping its twists and turns and charting its fluctuating depths. This river had been earmarked to be the primary artery for bringing in barges ferrying rigs to drill test boreholes in the forest but everything hinged on whether the river's width and depth allowed it.

Dammit. Without the camera I'd be missing a whole heap of pictures I was sure would be of interest to a photo library back home.

I wouldn't have been wrong. Almost everywhere one looked

during our river trips there was something worth snapping culminating in the most snappable event of all, a moment of near fatal death-defying madness involving those of our number all but press-ganged into tackling the veritable maelstrom foaming over the sandbar separating the river from the Atlantic which Shell said had to be depth-checked.

Up until that moment, there'd been no shortage of volunteers for the river survey job. All it would involve would be piloting a local pirogue – a long open canoe fashioned out of a substantial tree trunk – down the Manji River to the sea, recording its meanders with our faithful compasses and Topifils and noting down the depth readings measured with a mobile echo sounder.

Compared with our previous mission this was going to be easy street incarnate, I thought. No hacking through tangled viper-infested undergrowth or miles of debilitating mud-covered foot slogging, just sitting on our butts floating down a placid river where the only danger lay in dozing off in the heat and toppling into waters known to be inhabited by venomous water snakes and voracious crocs.

I could do that, I said, and I did.. for six blissful weeks.

While the rest of the crew continued their lagoon mapping activities, six of us were put on the river survey including Geordie, co-opted onto the job once his swamp-mapping mission was done with.

'Ah cannae believe it,' he enthused one night over a bottle of Cointreau consumed in the river survey camp we'd set up close to where one of the Manji River's tributaries joined a small lagoon near the sea. 'We're actually getting paid for this!'

I knew what he meant. Left alone by Shell and our respective employers to just get on with it, our little crew's daily routine was to rise early, motor across the lagoon to the beach in a Zodiac inflatable, battle with the surf to cure our hangovers, return for breakfast and, when we were ready, take to the pirogue for a few hours of river surveying before returning for dinner and creation of the next day's hangover courtesy of a well-stocked cocktail cabinet. There were those we knew back home who'd pay quids for this.

Did we feel any guilt over not busting a gut with this little exercise, over spanning the job out as long as we reasonably could to maximise the per diem payments coming our way for having fun in the sun? After what we'd variously been through to date, not exactly. Well, not if you didn't count Miewkeus' practise of selling our surplus, Shell-provided, beer and fuel supplies to willing buyers in villages along the river and ploughing the proceeds into keeping the cocktail cabinet topped up. Turning a blind eye to that did tweak the conscience a bit.

But not for long. The practise of our inexplicably-promoted to hydrographic camp manager madcap colleague was, as Geordie and I decided while reaching for another Cointreau top-up, a win-win situation for all concerned, even Shell which was paying for all this. The villagers would be getting heavily discounted beer and fuel, we'd be avoiding being poisoned by Gabon's curious concept of what constituted beer and as a result there'd be far less likelihood of illness interrupting the gathering of the data Shell needed to be able to decide whether the river was suitable for bringing in their well-drilling barges. What was not to like?

Well, one thing perhaps. The strong possibility of all this exploration work resulting in devastating consequences for the forest and everything that lived in it. That could lead to the odd sleepless night... especially if what we'd been hearing was true. That the area Shell had earmarked for exploration drilling and we'd been engaged in mapping for the last six months was, in fact, part of a designated forest protection zone.

Chapter 3.6

Oddly, it was Miewkeus who first raised the alarm. A man so detached from reality one often wondered how he'd ever come to be working for our militaristic mob let alone Shell.

It was because of his contacts, apparently. Miewkeus knew everyone hereabouts and OTC was prepared to put up with his bizarre behaviour so long as he continued to be able to make things happen. Things like keeping our local labour force from rioting or procuring items the rest of us would've had to sweat blood to obtain.

So all in all, hiring a near gorilla with virtually no English and even less connection to the civilised world to be the base camp gopher and subsequently river survey camp manager did make some sort of sense. Without him, Jacko calculated, everything we'd been hired to do would have taken at least twice as long.

On the other hand, that was balanced out by it taking twice as long for us on the river survey team to make our needs understood. Since everything had to be conducted in a mix of fractured French and the most basic of basic English, it wasn't unusual for requests for assistance to result in misunderstandings that could on occasion end up in both sides tearing at their hair in exasperation and in approaches to Jacko to please replace him with someone we didn't have to draw pictures for to get our requirement needs across.

In the end, after Miewkeus had disappeared for two days and returned with a new chef after our quite adequate one had got understandably irate over being unable to make him comprehend that the cooking gas cylinder was almost empty, Jacko had had to come to have a word. One that concluded with Miewkeus threatening that if Jacko so much as thought about replacing him as river survey camp manager he'd blow the whistle on how Shell had come to be exploring for oil in a forest protection zone in the first place.

Forest protection zone? Since when? No one had ever mentioned that to us. Was this just another Miewkeus miscommunication or was it a ruse to save his own skin? Without any means of checking the claim we couldn't tell. But surely it had to be an exaggeration. Shell couldn't be that stupid or arrogant, could it? Considering how much time and effort it put in trying to convince the world of its environmental credentials, news leaking out of it drilling for oil in such a sensitive area would be a public relations disaster.*

But the Gabonese/Swiss/South African remained resolute. Speaking more slowly and clearly than I'd ever heard him over the past five months, the man drummed the message home. He was adamant. This was exactly what Shell was doing and he'd only stayed mum about it because he needed the job. Take that away from him and…

He hadn't needed to finish the sentence. Even in the face of having only Miewkeus's garbled French Gabonese word for it, Jacko knew he couldn't take the risk. It being common knowledge that our base camp boss was angling for permanent job with Shell helping to manage the exploration drilling phase of the project, we all knew what his reaction would be. Miewkeus

---

* Enquiries made regarding this allegation in the course of writing this book produced a non-committal response from Shell. Although a number of maps unearthed during an internet search reveal the area concerned to be a designated forest protection zone (see Gabon map on page 124), the only comment a company spokesman felt able to make was that, since the allegation relates to Shell's onshore exploration activities conducted in Gabon in the 1970s, 'it's too far back for us to be able to reasonably provide details'.

could keep his job... providing he worked on his listening skills.

It was a fudge and everyone knew it. Listening was the last thing he was going to work on and the real message in the warning was aimed more at us than at Miewkeus. What Jacko was really saying was that it was down to us to be more clear in communicating our needs. Starve because there was no gas with which to cook the food and that'd be more our fault than the river survey camp manager's.

So which, we wondered, was the more dispensable to the project? Him or us? Us, it seemed, something that was to become all too evident the day we were ordered to put our lives on the line to collect a series of depth readings across the teeming sandbar separating the sluggish Manji river from the pounding Atlantic.

***

'Don't be such a wuss,' said Jacko on onpassing Shell's order to me. 'It's just a bit of water. You 'fraid of getting wet?'

Damn right. And of staying wet both inside and out. Go overboard in that maelstrom and even with a life jacket you stood only a slim chance of coming out still breathing. Without one or any rescue craft standing by to extract us – neither of which was being made available – your chances were so slim I doubted even a champion surfer trying to impress the girls would dare venture into that watery hell.

'OK then,' said Jacko after I'd told him this was a madness too far, even for someone who'd been putting his neck on the line every day for four solid months in the jungle, 'anyone here who DOES value his job enough to make sure the client stays happy? I need four volunteers.'

Four brainless twats with a death wish, I thought. One twat to man the outboard motor, one to run the echo sounder, one to take the readings and one to act as ballast to keep the Zodiac's bow down in the heaving surf.

'I'll start you off,' said Jacko to the surprise of no one, 'I'll drive.' Being seen to be the first to volunteer would do his chances of becoming a Shell employee no harm at all and everyone knew it. 'Who's with me?'

Knowing it wasn't so much a question as a scarcely-veiled threat to their future employment prospects, three others, all with minds pummelled into subservience through lengthy periods of servitude in Her Majesty's armed forces, raised a nervous finger hoping desperately there'd be more fingers in the air than volunteering positions. The three weren't the only ones with demanding families back home fully dependent on the money they sent back from this equatorial hellhole.

The look in their eyes on finding no other fingers raised said it all, especially to me, Geordie and my fellow Jungle Bunnies.

'This is what YOU signed up for, c**ts,' they said, 'not me. I just came to look after base camp's communications/electrics/water supply. YOU'RE the fucking surveyors. So what the fuck am I doing, doing your fucking job for you?'

Being a mindless twat with all the sense of the average lemming, my own eyes shot back. You might've been lobotomised into obeying orders without question regardless of the risks involved but there are some amongst us who were sadly denied that pleasure. Anyway, someone's got to stay behind to pick up the pieces and write letters of condolence to the victims' families. It's a dirty, difficult job, that we know, but we're more than willing to offer our services to undertake that onerous task.

As it turned out the only letter of condolence that did have to be written was the one Jacko himself was forced into writing informing Shell that the mission had fallen marginally short of its objectives. I didn't envy him having to find the words to express in euphemistic terms that the mission's only success lay in his crew of hopeless dimwits surviving being tossed around like a paper cup in a hurricane while failing utterly to gather any useable data on the sandbar's depth.

To give Jacko his due, his failure to fulfil the task in hand wasn't through lack of trying. Four times he'd powered the Zodiac into the maelstrom, managing to criss-cross it on three occasions as his crew struggled to carry out their respective tasks with one hand while hanging on for dear life with the other. Amazingly, in that respect the mission was a complete success. As later inspection of the echo sounding readings revealed, not only had they managed to keep the machine recording but each pass had produced several depth readings.

So far so good. But if Shell managed to make any sense of the readings, they'd be better data analysts than me or any of my surveyor colleagues. From the figures recorded while the craft was being bounced high and low in the frenzied surf you could end up concluding that the water depth over the sandbar was anything from almost zero to several metres, something any half wit could have predicted without having to put the lives of a third of the entire OTC complement in peril.

Watching them drag themselves ashore after finally deciding enough was enough I was reminded of the aftermath of the Charge of the Light Brigade in the Crimean War and the 'Futile Gesture' sketch in the 1960s comedy review *Beyond the Fringe*. Just to satisfy some lunatic order coming down from above, the boat crew had consigned all common sense to the trash can, saluted the flag, stiffened the upper lip and charged mindlessly into insurmountable odds gaining nothing in the process save for the mental derangement award conferred on them by a gaggle of local fishermen from the nearby village of Sette Cama standing open-mouthed and dumbstruck on the river bank.

As this mission of madness unfolded, the group's gradual crescendo of jabbering and gesticulating prompted a thought to cross my mind. If even the local fisherfolk wouldn't contemplate venturing into this maritime mayhem, what would the boat crew's trade union have made of their members being given no option but to undertake such an insane exercise?

***

Mulling it over for the remainder of the tour and, at the end of it, all the way back to the UK, just one likely union reaction remained after all others had been considered and summarily dismissed. Outrage. On learning that their members had been effectively told their jobs were on the line if they refused to go completely unprotected into the jaws of hell, the men's employer would be counting itself extremely lucky if it escaped severe retribution by both the trade union concerned and the courts. There were laws protecting worker rights through which this sort of coercion drove a veritable wagon train of carts and horses.

There was only one thing preventing that retribution happening, it occurred to me as we boarded the plane for a final spot of R&R in Port Gentil before taking our leave of Gabon. So far as I knew, none of the crew was a trade union member. Yes, in the event of being dismissed for refusing to carry out a clearly mad contractual requirement, they could of course launch a private action against OTC. But that would be both costly and time-consuming and, in the face of a raft of company lawyers opposing them, success for wrongful dismissal would be by no means guaranteed.

So didn't it make sense to join a union, I asked Geordie by way of taking his crazed mind off the first white women we'd seen in months cavorting half-naked on the PG beach? For all we knew, letters from our respective employers might be waiting for us in the UK informing us we were now surplus to requirements, our employers' way of telling us we were being made examples of for failing to cooperate in the completion of the survey mission to Shell's satisfaction.

'Ah dinnae care!' mumbled Geordie into the sand, unable to turn from lying on his front to hide a monstrous and uncontrollable erection. 'Just get me one o' them and ah'd walk through fire if tha company asked me!'

Fair point. Finding ourselves in the midst of a bevy of curvaceous French expat wives and daughters after being denied any female company for so long was a bit distracting, and although we didn't know it at the time, things for poor Geordie in particular were only going to get worse.

Finally taking our seats on the plane out of Libreville airport, en route to Paris and then London, I sensed the man beginning to groan in libido frustrated despair. After the PG beach experience, the last thing he'd needed was to find our path to the departure gate blocked by a phalanx of schoolgirls dressed in what looked like Girl Guide uniforms, every one of them clutching a Kalashnikov semi-automatic rifle to their ample bosoms, there, apparently, to act as a guard of honour as the president's plane came in.

'Oh Christ!' he whined in tune to the aircraft engines. 'Won't someone PLEASE have pity? Was that someone's idea of a sick joke, making flesh every fantasy I've ever had then denying a bloke the chance to fulfil them? This is sheer bloody torture!'

I knew what he meant. We'd all had our unfulfillable sexual fantasies over the months, something that'd driven every one of the crew to the only remedial medication available, the camp cocktail cabinet. Heavy application of the medicines contained therein had helped us through it then, so, I suggested to my now near frantic travel companion, now might be an appropriate moment to consider a repeat prescription.

Geordie was in no position to demur and by the time we landed at Heathrow some twelve hours later there was more alcohol in the pair of us than was left in the plane's booze locker, a condition that did not go unnoticed by the customs man as we weaved our inebriate way through the 'Nothing To Declare' channel.

'All right lads?' he smiled at us. 'Like to take a squint at your bags if that's OK. Looks like you've been away a while. Anywhere nice?'

'Africa,' I blurted out before I could stop myself. If there was one place that'd win you the full bag, possibly full body, search it was the dark continent, well known for being the source of some of the most potent substances on HM Customs & Excise illicit drugs list.

'That so? Well, hope you haven't packed anything venomous by mistake. Better take a look, lads, just to be on the safe side. Any nasty surprises you'd like to make me aware of before I do?'

Can't think of anything, our inebriate shakes of the head were intended to inform him. I didn't know about Geordie but so far as I knew I was clean. Well, I was if the man didn't count – or find – the bits of crocodile skull stumbled on on the banks of the Manji river, subsequently boiled and bagged and stashed away in the depths of my backpack until I had a chance to glue them all back together.

I wasn't sure what the man did expect to find but it clearly wasn't that. Response – a single, very raised eyebrow and a little crooked smile that told me I was now very much for it.

'Interesting,' he said on shaking the bits of skull out onto the work surface. 'Haven't seen one of those for a while. Anything else you think I might be interested in? How about this?'

Oh bugger. I'd completely forgotten about that, the rhinoceros beetle I'd saved from doing its species' usual trick of flying into the campfire and going bang. As entertaining as that was, I'd decided to capture one for the Africa bits and pieces collection and had entombed it in a tobacco tin filled with kerosene in the hope that that would act as a preservative.

Had it been just a tobacco tin, it might not have attracted much attention. But being sealed with several layers of insulating tape to prevent leakage made it an instant item of interest and now our customs man friend was frowningly engaged in de-sealing it.

'What have we here then?' he said on removing the last bit of tape.

'Rhinoceros beetle,' I slurred back as he started to unscrew the top.

'Careful! It's full of...'

Too late. He already had the top off... to reveal a heap of green powder where the beetle used to be.

'Where?' he said, showing me the contents. 'Looks to me like your bird has flown,' he smiled, dipping his finger into the powder and giving it a lick.

At first he remained smilingly pensive, allowing his taste buds time to conduct their assessment of the new flavour in their presence.

Then his smile evaporated, his face changed to the same shade as the powder and without so much as a fare thee well he dropped the tin and fled behind a screen leaving Geordie and me wondering how we should be reacting to the violent retching noises the screen was now emitting.

Our first thought was to call for help. But as there seemed no one to call for help to, option two was quickly selected. The one which saw us hurriedly stuff everything back in our bags and scurry for the exit. He could keep the tin.

'I hope he'll be all right,' I said to Geordie as we ran. 'Don't want him on my conscience as well as everything else.'

'Dinnae fret yoursel', bonnie lad,' said Geordie with drunken grin. 'He'll be fine and we've got pussy to chase.'

Chapter 3.7

Poor Geordie. If he thought getting back to home shores was immediately going to solve his libido problem and help him forget being the recipient of the Dear John letter all of us dreaded, he was going to be disappointed. The 'business' ladies occupying stools in the bar of the snazzy airport hotel we'd booked into proved to be more discriminating than expected and by the time he'd been rebuffed by every one and then persuaded by the musclebound barman that it was long past our bedtime, the poor lad had reached the stage of such frantic desperation he'd even started eyeing-up the black-clad, wheelchair-bound dowager being attended to by a surly African manservant in the lobby.

Dragging him away and into the lift in a futile attempt to preserve what little was left of his dignity I tried to explain the reason for his lack of success in trying to attract a mate for the night.

It had nothing to do with what he'd begun to think was some antimalarial drug-caused decline in his masculine appeal, more likely the exact opposite. The barstool 'ladies' in such an upmarket place would likely be saving themselves for their regular clientele of business executives with generous expense accounts and time to kill while awaiting connecting flights. Even at the peak of the season of giving, giving out to a couple of leering drunken gorillas in crumpled shirts with their tongues hanging out would be a gift too far while there was still the chance of hooking a client likely to make less demands on them than those giving the impression of having just escaped from a lengthy spell of female-deprived captivity.

'Christ! Is that really what we look like?' Geordie mumbled into his cup of tepid khaki-coloured fluid the hotel had the gall to call coffee at the breakfast table the next morning. 'Creatures from the black lagoon? No wonder we didn't get laid.'

'Speak for yourself,' I mumbled back.

'You mean YOU got laid?'

'Nah. Creature from the black lagoon. I'm more something that crawled out from under a rock if the mirror was to be believed this morning. Anyway, it wasn't so important for me to get my leg over as it was for you. Just a few more hours and I'll be getting all the comforts of home from Flame Hair. Well, providing she hasn't left town since I last saw her.'

'Lucky bastard. Know what I've got waiting for me after that letter? Nothing. Nada. Zilch. Zero. And all in a place that'll be colder'n a witch's tit this time of year. Might as well not have left the jungle. At least there there'd be the chance of having a nice warm black widow to snuggle up to.'

'Speaking of which...' I gestured across the room towards the dowager in the wheelchair who'd just been wheeled in by her African slave. 'Hang around a while and you might get a chance to have a second go at it.'

'Nah. Think I'll give that a miss. Even waking up in a house as cheery as the grave would beat waking up next to a cadaver. Did I tell you me da's been laid off from the steelworks? Means Christmas in our house'll be about as jolly as losing to bloody Sunderland.'

Jeez. Bad as that, eh? I'd forgotten that the people of The North would be being hit twice as hard by the economic slump as us softies in The South and that at such times the likes of Geordie would look for solace in their football team's performances.

So losing to Sunderland, my God, that didn't bear thinking about in the Newcastle-supporting Geordie household and, as we parted in central London, him heading north, me south, it was as much as I could do to prevent the relief of heading in a different direction showing. Poor bloody Geordie. Fancy being faced with having to go somewhere that'd actually envy places tagged with the dark satanic mills label. At least in places with such a mill still operating there'd still be the chance of a dark satanic job.

***

As relieved as I was to be travelling away from such desperate places, a question still nagged all the way to my parents' house in Brighton, my between jobs 'hotel' and mailing address. After declining to offer my services for the Manji river sandbar experience, would I still have a job, satanic or not? By the look on Jacko's face on his return from his little boating trip, if I did it'd only be because Jacko had himself been fired for presenting Shell with a sheaf of utterly unusable sandbar depth readings and within moments of stepping through the front door I was leafing through the mail almost certain there'd be a suitcase letter waiting for me.

Phew. Nothing. For the moment at least I was still in full employment, a finding that had me heading straight to the pub to celebrate with friends after re-introducing myself to my mother who hadn't recognised the grubby degenerate who'd somehow got hold of a key to the house.

The hope was to surprise those not knowing I was back, as in everyone. In my haste to leave, I'd forgotten to let anyone know the mission was over and I was heading home.

It was a surprise all right, just not for them. For as I edged unseen into a pub crammed with pre-Christmas revellers the first 'friend' I set eyes on was Flame Hair, hunkered down in a corner with the lead guitarist of a band I knew she followed and clearly having no objection to having her neck nuzzled.

Frozen in mid-step, two thoughts surged through my head. One, that staying to see how things panned out was the last thing I wanted to do. Two, that if Geordie had seen what I'd just seen it would've made his day. Able to spot insincerity from the other side of the impenetrable forest we'd first met in, he knew as well as me that all that consoling talk and pats on the back over being the recipient of a Dear John was about as heartfelt as the devastation I'd expressed on lashing blindly out in the middle of the night to flatten a couple of frogs in mid-copulation on the tent pole next to my ear. Both expressions of devastated concern hid a common factor of self-interest. In the latter case, now I'd be able to get some sleep, in the former, thank fuck it's you getting the letter, not me.

'OK, pal,' I could hear him intoning in my head as I turned to retrace my dazed path back out of the pub and on to the street. 'Get it now? What goes around comes around, bonnie lad, more often than not compounded up. At least my ex was considerate enough to let me know it was all over. Yours? Well, frankly I reckon I'd prefer a Dear John to getting back to find I was the last to know what games the mouse had been playing while the cat was away.'

Bloody Geordie, I thought as I carried a stiff drink to an empty corner table in a pub both I and my friends rarely used. Did he have a point? Deep down, was Flame Hair THAT untrustworthy? THAT inconsiderate towards my feelings? Maybe. But on the other hand, could I really object to her having a bit of fun in my absence? Reverse the roles and I'm not sure I'd have been able to stay faithful to any vow of celibacy for the length of time I was away. In fact, I know I wouldn't. I'd proved it on more than one occasion while on leave in both Ethiopia and Gabon.

Still, I thought, gazing blindly into a glass that seemed to now be looking accusingly back at me, it would've been nice if she hadn't got my hopes up…

Chapter 3.8

'Wassail, wassail!' slurred the man bedecked in streamers and glitter who'd lurched into the table spilling both my drink and his over me as he staggered to regain his balance. 'Oops, sorry boyo. Lemme get you a refill. Holy fuck. It's you! Well bugger me backwards 'til Sunday!'

Oh great. HJ. I didn't deserve this. Must be back from Libya or wherever he was now selling his oily well-logging soul. Back for Christmas with the thirst of a dying dingo and the cash to try to revive it. The absolutely last person I wanted to spent this particular evening with.

'Jayzus. Will you look at you, boyo? Bit of a change from last we met. Get a promotion to deckchair attendant-in-chief, did you?'

Slumped grinning and trying to focus on me from the chair I'd been hoping would remain empty for the evening, he made it clear there'd be no easy escape from him this night. No matter how I tried I could be sure he'd be dogging my every step until I gave in and agreed to swap tales of the great British/Welsh migration to the only country that'd have us, the land of black gold and even blacker hearts.

'Follow my advice then did you, boyo?' Even in his most inebriate moments I knew from college days HJ could recall every word he'd ever uttered and even years under the Saharan sun seemed not to have dimmed his photographic memory.

Bugger. I was hoping it had. Now I'd have to recount how that advice, received what now seemed like aeons since, had led to my own descent into a life of unprincipled rape of the resources of the most under-developed, most resilience-incapacitated nations of the world.

'Yeah, great innit?' he beamed as I mumbled my way through a much abbreviated summary of the last few years. 'It's neo-colonialism all over again, boyo. Just what such places have been crying out for. Independence? Don't make me laugh. They're about as independent as you and me. Take the oil industry away from them and they'd be back to shagging camels in days. Oil's all they've got and they know it. And if it wasn't for the likes of you and me, it'd never get out of the ground and into the coffers that'll one day bring them actual proper houses to live in.

'So don't know about you, boyo, but no regrets there. No cracks in this boyo's conscience. Not while the beer tokens keep pouring in to keep me from dying of thirst. Talking of which, your shout I think.'

Bugger again. Couldn't let him get away with that provocative little soliloquy. The man had thrown down a challenge to my own, more considered view of western imperialism and he somehow had to be kept in place to hear it. There was only one way. Drag my heels over buying my round. Never known to walk away while a drink was owed him, there'd be no problem retaining his undivided attention while his glass remained unreplenished.

'So... no cracks in the conscience, eh?' I said while lingering over finishing my own drink. 'Ever been in the jungle?'

'Noooo, boyo. Strictly desert for me. Nice clean place with ne'er a tree to spoil the view of the rigs pumping my lovely salary out of the ground. Just the way I like it. Jungle's for rejects, boyo. No offence.'

'So you won't have seen the damage the drilling does to it,' I countered trying to ignore his deliberate sideswipe at those he clearly viewed as the oil industry's lesser orders and thinking back on Miewkeus's drilling for oil in a forest protection zone revelation. 'If you had,' I said, 'you might not be able to sleep so soundly at night.'

'Doubt it, boyo. Bloody big world out there. More than enough room for both a bit of drilling and your precious jungle. So what if a few trees go missing? Plenty won't and the benefits of cutting a few down to allow us to drill more than compensates for their loss. The world needs oil, boyo, and it'll pay a pretty penny to get it. So it's win-win for everyone, I'd say. The world gets the oil it needs, the oil producing countries get a heap of money they wouldn't be getting if we weren't there helping them get at their resources and my lovely salary keeps rolling in. What's not to like?'

Dammit. HJ had a point there. It was a bloody big world and when you thought about how little of it was actually needed to accommodate the rigs and refineries and ancillary stuff, what were a few trees in the context of things? It seemed like a checkmate point.

Or it did until it dawned on me that this line of thought was in fact a double-edged sword. If there was so much room out there for a few oil exploration and production operations, why would anyone need to go exploring in a protected area? Surely the likes of Shell – if it was true they had had us working in one without letting on – had enough concession areas to leave such bits of forest alone, not to mention everything that lived in them.

'Ours not to reason why, boyo. That way madness lies, as the bard would say. Start down the spiral of questioning your lord and master's reasoning and you could well end up spiralling back up your own fundament. Take it from me, bach. I've seen it happen. Better men than you and me have ended up worrying themselves out of a job with such fretting.'

'What happened to them?'

'Back on the scrapheap, bach. That what you want happening to you?'

Not especially. But it'd be difficult turning a blind eye to such an illicit activity if the drilling in a protected area allegation turned out to be true. Quite apart from the forest destruction factor there was the little matter of the deceit involved. Making me an unwitting accessory to a crime I had no idea was in progress wasn't the best way I could think of for engendering undying commitment to the oil world.

Neither, for that matter, was the Manji river sandbar 'expedition'. If they thought making me and my crew feel so expendable would endear us to them, they were even more crass and arrogant than I'd thought. All that order had achieved was to prompt some of us to start thinking seriously about joining a union. And while we're on that subject, I put it to HJ, if more of us were unionised that'd make them think twice before being so cavalier with their workers, surely.

'Wouldn't bank on it, boyo. Know why none of your crew mates is in a union? It's cos the oil companies make sure no one who is gets anywhere near working for them. Makes a lot of sense when you think about it. The last thing they want is anyone getting bolshie. Bad for production. So in effect, unionisation is outlawed.'

'And you can still work in such an industry? A boy from the valleys like you? What would your ma and pa have to say about that, good mineworker union folk that I expect they are. Don't they nag you about how the miners would never have won better pay and conditions had it not been for the union?'

'Bloody miners! Don't go sticking them in my face. Let me tell you about the miners, bach. They're as bad as the bloody oil barons. Both closed shops the pair of 'em. It's just that one shuts you out if you are a union member, one shuts you out if you're not. Same same with mining families. That's why I'm down here for Christmas, not there. Not being in a union I'm all but blacklegged from mine.'

Interesting. This wasn't the first time I'd heard of family pressures having a bearing on employment prospects in the energy industry. The only difference was that on the other occasion the exact same sentiment had been voiced by a Shell engineer at the Gamba oil terminal.

After having had his tongue loosened with more drinks than was good for either of us I found myself being treated to a scarcely believable account of how the family aspect also featured in the oil industry's standard hiring procedure.

Following years of knocking on Shell's door, the engineer had finally got an interview which had gone quite well. Yes, said Shell, they could offer him a job. At the Gamba oil terminal. Just one thing. Could they also talk to his wife, just to be sure she was under no illusion as to what living in West Africa would involve.

OK, said my friend. I'll ask her to drop by.

'And ask her to bring the children,' they said. 'We'd like to meet them too.'

Holy crap. And if he had reservations about subjecting his kids to a Shell 'interview'?

'Then although they didn't say so in so many words, what they were telling me was that it might take a while for them to confirm the job offer. In effect, they were saying it wasn't just me they wanted. It was a family unit. One that slotted nicely in with the Shell profile.'

Very much like the miners and their union I thought as HJ's own story unfolded.

Christ. What a parallel.

Chapter 3.9

Although I didn't know it at the time, I was about to discover that I too was far from immune to family pressures in relation to my own career path. Pressures exerted on me the following morning in the form of 'advice' from my mother regarding the extra curricular activities associated with the work of the exploration surveyor.

Waking up in a place I didn't recall going to bed in after somehow making it home from my 'discussion' with HJ, the first thing that struck me was that I didn't remember going to bed sitting up. Neither did I recollect getting every stitch of clothing off. But both things now applied and so did being covered not in bedclothes but in the dog's smelly blanket. If the evidence was to be believed, everything pointed to my having been so incapacitated I hadn't been able to make it further than a chair in the living room.

Not so, my mother informed me on noting a stirring in the chair. I had made it to my own bed. She'd been woken by the noise of me crashing in through the front door and stumbling up the stairs.

But she'd also been woken in the early hours by the sound of someone struggling to open the back door. Peeking through her bedroom curtains to catch a glimpse of whoever was responsible she'd been surprised to find the moonlight illuminating not, as she'd thought, an axe-wielding madman but the son who, until his surprise appearance the previous night, she'd thought was still somewhere in Africa. A son that was now engaged in staggering haphazardly across the lawn towards the less well-kept jungly part of the garden dressed only in the shirt he'd gone out in and clutching a roll of toilet paper in his hand.

Falling flat on his face over a garden chair in his path, he'd then proceeded to clamber onto the chair, vomit voluminously down his shirt front and pass out.

At first, my mother expounded, she'd been tempted to leave him there to learn a very chilly lesson. But then her humanitarian nursing conscience had kicked in and she relented. On letting the dog out the next morning, the last thing she wanted to find was her son's lifeless body fixed frozen to the chair. It was quite new and in the current economic climate she knew my father – who'd slept soundly throughout – would raise Kane on being asked to fork out for a replacement.

So she'd donned dressing gown and gum boots and ventured out into the cold to see what could be done. In the end, there was nothing else for it but to heave her wayward son to his feet, lug him back indoors, remove his vomit-soaked shirt, deposit him in a chair and cover him with the first thing that came to hand. Yes, the dog had complained a bit about being deprived of its nice warm smelly blanket but that was a small price to pay for making sure she was protected from the sight of the boy's unadorned full frontal nakedness when she came back down in the morning. Not the best way to start the day.

The lecture I received after being appraised of the situation came in one single look. 'Are you sure your current choice of career is the best one for you, dear?' it said. 'If a life of being deprived of civilised surroundings for lengthy periods is having the effect of making you forget your social skills, might it not be better to transfer to some less challenging indoor occupation? Far be it for me to live your life for you, dear, but you might find such a transfer more amenable to outliving your parents.'

Oh God. She could well have a point there I thought as I clambered up the stairs to take refuge in my room and bed. Was this life having the dehumanising effect on me her eyes indicated it was? Being found in the garden jungle in the middle of the night with a toilet roll in my hand suggested there was no doubt. Such subconscious behaviour was now so ingrained in me I wondered if it could ever be eradicated, 'sensible' indoor job or not.

So where did that leave me? Just give in to it and play the cards I'd been dealt or chuck in my hand and ask for a re-deal? Both carried similar risks. There was no guarantee a fresh hand would be any better.

But on the other hand, what choice did I have? Sticking with what I had wouldn't just see me sink inevitably deeper into the abyss of oil industry dehumanisation but would, in effect, be condoning the practises of a body of people I'd begun to despise as much as a world that was doing nothing to rein in such practises.

Thinking about it, this wasn't the first time I'd wrestled with this imponderable. I'd been here before, a lot longer ago than I'd previously been prepared to admit. If I was being completely honest with myself such qualms went all the way back to the day I first arrived at the 3S camp in the Ogaden. The discovery that all that company recruiter talk of being posted to the oil industry's equivalent of an all mod cons holiday camp was as honourable as a Tory Prime Minister's social conscience had started the removal of scales from previously starstruck eyes and one by one they'd continued to fall until the last two peeled off just days before my departure from Gabon.

At first I'd thought the Manji river sandbar incident had been the moment the full, unadulterated truth about the industry I'd shackled myself to had been revealed. Nothing else got close to exposing the oil world's true attitude towards those caught in its web.

But then I'd thought back to Miewkeus's forest protection zone allegation and revised my thinking. THAT was the real moment I'd seen the light. Even if the jungle I'd been hacking my way through wasn't a designated forest protection area it bloody well ought to be and any organisation with a conscience would not only have seen that but would've had severe reservations about disturbing what in so many respects was a unique part of the natural world.

That it hadn't, told me everything I needed to know about Shell and its like and about my future career path. Even if they could sleep soundly at night knowing they were destroying something that could never be replaced, I couldn't and that left me with just one option. The one that would see mother getting her wish... or something very much like it.

Before I found myself sucked inextricably down into this unconscionable world a suitable exit strategy had not only to be found but implemented. A strategy which would need to do two things at the same time – get me out of this despicable industry while not at the same time threaten a future in which the skills learned at college and beyond could be gainfully employed. As honourable as being a deckchair attendant was, there was some doubt in my mind as to it being an altogether fulfilling career for someone who'd proved he could do an astrofix calculation while wresting monster pythons in the heart of darkest Africa.

But what career would be? That was the question that occupied me for the rest of my leave, and as the days ticked away, a mapmaker's plot for navigating a path out of this latest jungle gradually emerged. One I sincerely hoped would keep every side of every party happy.

Part Four

SAUDI ARABIA 1976

What goes around comes around

Israel
Syria
Tel Aviv-Yafo
Jerusalem
AMMAN
Jordan
BAGHDAD
Iraq
Karbala'
Al Hillah
An Najaf
Al 'Amarah
Al Basrah
Abadan
Ahvaz
Iran
Kuwait
KUWAIT
Turayf
An Nabk
'Ar'ar
Ma'an
Elat
Al 'Aqabah
Al Jawf
Sakakah
Rafha'
Al Bi'r
Tabuk
Al Qalibah
Tayma'
Jubbah
Baq'a'
Ha'il
Hafar al Batin
King Khalid Military City
Ra's al Khafji
Bandar-e Bushehr
Shiraz
Al Ghardaqah
Bur Safajah
Duba
Al Wajh
Umm Lajj
Al Jubayl
Ras Tanura
Al Qatif
Ad Dammam
Dhahran
Bahrain
MANAMA
Persian Gulf
Buraydah
Az Zilfi
'Unayzah
Al Majma'ah
Rumah
Najd
Al Hufuf
Qatar
DOHA
Dubayy
ABU DHABI
Medina
Ad Dawadimi
RIYADH
Al Kharj (oasis)
As Sulaymaniyah
As Salwa
Harad
United Arab Emirates
Yanbu' al Bahr
Yanbu' al Sinaiyah
Badr Hunayn
Tropic of Cancer
'Afif
Al Quway'iyah
Halaban
Red Sea
Rabigh
Zalim
Saudi Arabia
Jiddah
Mecca
At Ta'if
Ar Rawdah
Al Khamasin
As Sulayyil
Rub' al Khali
Oman
Port Sudan
Al Lith
Qal'at Bishah
Sudan
Al Qunfudhah
Khamis Mushayt
Abha
Najran
Ash Sharawrah
Wudy'ah
JAZA'IR FARASAN
Jizan
Sa'dah
Salalah
Kassala
Ak'ordat
Asmara
SANAA
Yemen
Ethiopia
Al Hudaydah
Sayun
Al Mukalla
Ash Shihr
Sayhut
Arabian Sea
SCALE 1:12,500,000
100 200 300 400 Kilometers
100 200 300 400 Miles
Lambert Conformal Conic Projection, standard parallels 12°N and 38°N
International boundary
National capital
Railroad
Expressway
Road

Al Wajh
Umm Lajj
Hejaz
Buraydah
Az Zilfi
'Unayzah
Al Majma'ah
Rumah
Ras Tanura
Ad Dammam
Dhahran
Al Hufuf
Najd
Medina
Ad Dawadimi
RIYADH
As Salwa
Al Kharj (oasis)
As Sulaymaniyah
Harad
Yanbu' al Bahr
Yanbu' al Sinaiyah
Badr Hunayn
Tropic of Cancer
'Afif
Al Quway'iyah
Halaban
Red Sea
Rabigh
Zalim
Saudi Arabia
Jiddah
Mecca
At Ta'if
Ar Rawdah
Al Khamasin
As Sulayyil
Al Lith
Qal'at Bishah
Port Sudan
Al Qunfudhah
Rub' a

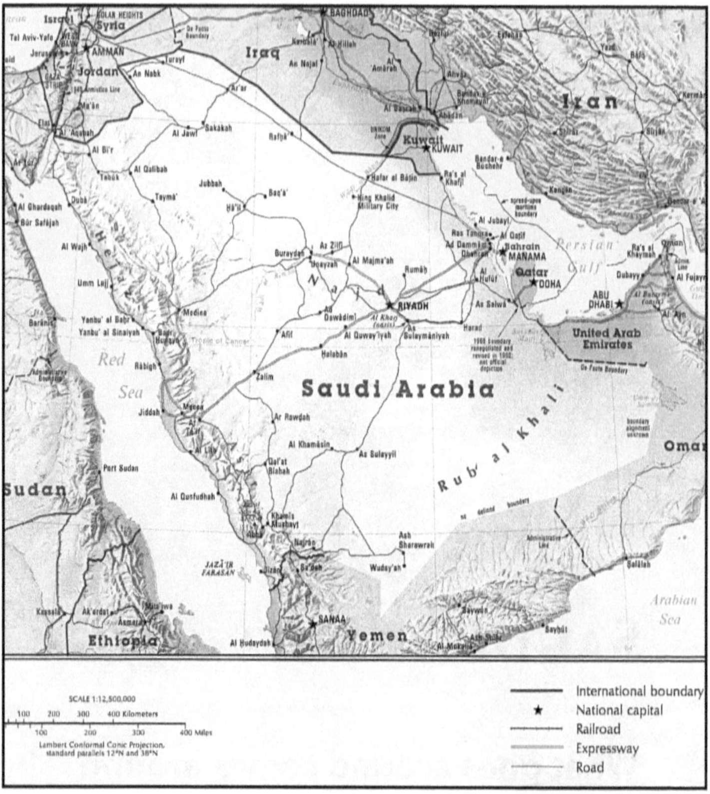

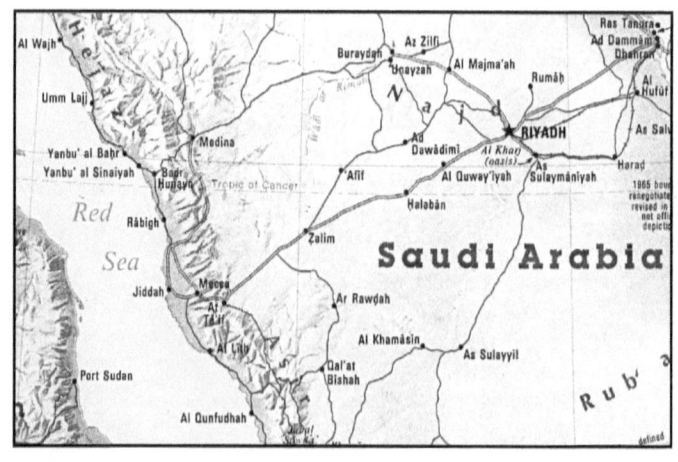

Chapter 4.1

Staring open-mouthed after the supply truck as it disappeared up the track in a cloud of desert dust I wondered how I could ever have been so stupid. I'd taken the OTC survey manager at his word and this was the result. A garden path that thanks to an unintended glimpse of the contents of my recent visitor's briefcase I now found I'd been well and truly led up. This was not the non oil world-related job I'd been assured I'd be on.

Thinking back on that meeting with the OTC survey manager I kicked myself for not seeing it. Both he and the company were duplicity incarnate. From bitter experience in Gabon, of that I was fully aware. So why hadn't I delved deeper on receiving his almost casual assurance that my wish to never again be put on an oil industry-related job would be granted?

The reason was obvious when I thought about it. Having seen how the economic slump had taken the colour out of even Christmas, the relief of knowing I'd escaped a return to the scrapheap of long term unemployment had trumped every other thought in my head. Such an outcome wasn't just possible, it was probable after having had the temerity to even hint at there being a limit to the type of job I could now accept.

Even as I opened my mouth at OTC's London HQ to make my request/demand I knew it was a mighty gamble. Oil was where the money was and the company that employed me was nothing if not avaricious. But despite that, something drove me on. A hunch that such a request might not be received in quite such the negative light I feared, possibly even the reverse.

How so? Well, after thinking about almost nothing else since pouring myself off the plane from Gabon the thought had finally hit me that my employer might actually welcome my request. With the oil industry mopping up so much of the world's survey-skilled manpower, wouldn't that be leaving OTC unable to accommodate other sectors into its programme of work? With my help, now it could.

Without figures or inside information to go on I had no idea whether that supposition held water, but instinct told me my gamble might not carry quite the risk I'd initially assumed. So in the end I'd decided to just go for it, holding my breath as my manager's brow furrowed in deep contemplation over being presented with such an audacious proposition.

With face about to turn blue, I'd started thinking I'd over-played my hand until he finally responded, putting me out of my misery with one brief managerial grunt.

'OK, you have your wish,' the grunt translated to. 'There is a non-oil job I can put you on. Be ready to leave in a week. Anything else?'

Well, a defibrillator to get my heart beating again, I thought. The stress of that moment could compare with anything the Gabonese jungle had been able to throw at me and left me wending an unsteady path out of the manager's office, knees the consistency of an over-milked blancmange.

Looking back on that moment two weeks later after settling into a survey camp in a remote region of Saudi Arabia, it occurred to me that, actually, there might have been more to the outcome of that 'interview' than I thought, notably the 'education' I'd received during card sessions with fellow desperados in like camps these past couple of years.

Had such sessions taught me how to keep a straight face when playing a poor hand? If so, maybe the card games hadn't been such an abuse of my time as I'd thought. If they had had an effect on the survey manager's decision to grant me my wish, shouldn't I actually be congratulating myself for my perspicacity in taking part in them? OK, I'd had to accept a modest cut in salary to be assigned to the Saudi job but that was a small price to pay for avoiding once again finding myself engaged in hunting for oil. So what, really, was not to like?

Well, perhaps that this survey of a barren stretch of desert was for some unspecified industrial complex and no industrial complex in Saudi Arabia would ever get built without the country's immense oil wealth, but at least there'd be no direct connection to the oil world and in the circumstances that, I decided on joining the survey crew in the mess tent for my first night in the desert for many moons, could be considered a very definite win.

***

The feeling of contentment over having succeeded in escaping the insidious oil industry's clutches stayed with me right up to the moment I opened the mess tent fridge looking for chilled lubrication.

Finding the cupboard proverbially bare save for the regulation Fanta and Sprite, the realisation of what I'd done finally hit me. I'd condemned myself to six months of a drinks diet consisting of anything I fancied bar anything that didn't fall into the soft drinks category.

Oh dear God. It was true. Saudi really was as dry, in every sense of the word, as they said and as the full horror of what I'd let myself in for sank in, from nowhere the sound of my mother's voice began reverberating gleefully in the back of my head.

'Good,' I could hear her beaming. 'Might keep you out of the cirrhosis ward.'

She hadn't needed to say more. I already knew her views on the subject of excessive drinking, something the garden incident hadn't done a lot to mitigate. Her eyes had told me at the time that if that was a reliable indicator of the sort of human being close association with the oil industry produced, that industry had a lot more to answer for than simply being the cause of a global economic collapse. It was also part responsible for the grief of mothers left weeping at the graveside of their over-alcoholled offspring and anything that could prevent her from joining that number had to be celebrated.

Even, I could hear her intoning in my head, the dispatch of this particular offspring to a country that had no hesitation in lopping the heads off anyone caught flouting Saudi Islamic law. Such a risk would be a small price to pay for keeping me from premature liver damage/death and anything she could do to prevent that happening she'd gladly do, even to the extent of helping buy my ticket to Saudi if that's what it took.

It wasn't, of course. The company had that covered. However, I might have welcomed my mother's input to help fund a transfer to an airline other than Saudia, the country's state-run airline that was as dry as the country I was travelling to. Flying without a beer or two to help me sleep through it was as disconcerting as knowing I wouldn't be tasting a bacon sandwich until I was safely back on home soil.

'Don't worry,' a friend's father who'd done time in Saudi assured me. 'While you're right about all pork products being banned, I'm sure you'll be relieved to hear you can still get a beer in Saudi. Although it is officially outlawed, foreigners can indulge in certain places without too much fear of retribution and Saudia's cabin crew usually turn a blind eye to a bit of booze in your carry-on baggage. So I wouldn't be too bothered about grabbing something from duty-free before boarding the flight. Just don't flout it when you're on board.'

I didn't need to be told twice. The relief of knowing I wasn't going to be made forcibly teetotal for the duration was palpable and the duty-free shop was duly visited prior to boarding… somewhat to the consternation of the sales assistant who checked boarding passes at the store checkout.

'Riyadh?' she exclaimed. 'You sure you want to take a bottle of vodka there?'

'No problem,' I assured her. I had inside information that a small amount wouldn't result in me being frog-marched to Chop Square, the westerner nickname for what was reputed to be public execution central.

'Well, on your head be it,' she said somewhat ominously, handing me my change.

My head indeed. One that was swimming in vodka and orange juice by the time we landed at the Saudi capital Riyadh, a condition that did not escape the customs official's attention.

'And this is?' he demanded, indicating with pointed disgust the half-empty bottle of vodka in my hand luggage.

'Remains of my duty-free,' I replied with as much of a look of innocence as I could muster. 'If anyone asks,' my friend's father had advised, 'just play the dim-witted ignorant tourist. It does a Saudi official's sense of superiority wonders finding himself in a position to help bring these stupid overfed westerners out of their educational ignorance.'

I'm not sure what I did expect this particular Saudi official's reaction to be but it wasn't what subsequently transpired. Rather than simply confiscating the bottle and sending me on my way with a flea in my ear, he took the bottle in one hand, my elbow in the other and started leading me across the customs hall to a door on the far side.

In that short journey about fifty possible outcomes to this scenario cascaded through my partly addled head, ranging anywhere from finding myself banged up in a holding pen to await the arrival of the Chop Square transfer bus to being the guest of honour at a party of his mates, all gagging for a dram of something a tad more invigorating than Arab mint tea.

The eventual outcome was rather more mundane than that. Pushed firmly through the door I found myself in a gentleman's toilet complete with western-style urinal. Ushered to stand in front of it, my customs friend looked me straight in the eyes as he unscrewed the bottle top, grabbed my elbow again and proceeded to tip the remains of the vodka slowly and deliberately down the pan.

'Get it?' his eyes said, fingers tightening vice-like on my arm as he led me back to my luggage and thence to the customs hall exit, waving me through and away with an exaggerated flick of a boney dismissive hand.

'Message received,' I murmured back, quickening my pace to escape into the night before he changed his mind and pointed me out to the ranks of severe-faced policemen pacing the hall beyond the arrivals gate.

***

'You did WHAT?' gasped the meeting committee of one on being appraised of the events of the past half hour. 'You DO know where you are, I suppose?'

'Yes, but someone I know who's 'done' Saudi said it'd be OK.'

'You sure he said Saudi Arabia?' said 'Genghis' Carne, OTC's local survey chief who'd dispatched himself to meet me at the airport and ferry me to the company staff house. 'Couldn't have been one of the more liberal Arab states, could it?'

'No. Definitely Saudi. Said he'd done time here a couple of years ago so I had no reason to doubt him.'

'Where in Saudi, exactly? Riyadh?'

'No. Somewhere beginning in 'D' I think he said.'

'Dhahran?'

'Yes, that sounds like it.'

'Ah well, that explains everything.'

'Er, OK… how?'

'Not really Saudi Arabia, Dhahran. Law unto itself. In effect an American enclave here. Aramco's big base.'

'Who?'

'Jayzus. You never heard of Aramco? Arabian American Oil Company. Biggest oil producer in the world. And if your mate worked there, not surprised he gave you such duff information. The place is like a small American town complete with nightclubs, bars, burger joints, drive-throughs and the like. Nothing you can't get there… including booze.

'Bit different here, but you've already discovered that. Don't want to worry you but had a stricter customs man been on duty what you did could have ended a bloody sight worse. They bang people up for just being in a minor car crash here. Especially if it's a Saudi you crash into. You might want to bear that in mind when you're driving down to Jeddah.'

'Where?'

'Oh for fuck's sake, didn't they tell you anything back in London?'

'Only that the job was some sort of survey of a bit of desert scrub.'

'Well at least they got that right. We've been asked to survey a site at a little place called Yanbu, north of Jeddah on the Red Sea coast. That's Saudi's second city after Riyadh, for your information, and where you'll be going once we've got you sorted out.'

'Sorted out?'

'Yeah. Work permits and stuff. Won't take long. Got a fixer here who does all that for us. Real Saudi chancer he is. Charges through the nose for the service but gets the job done. Couldn't operate without him. But suggest you don't mention the vodka thing to him. Not if you don't want to end up following the booze down the pan. For your information, you're in a city that's effectively the Islamic law centre of the world and bans more things than the Catholic church. Step out of line here and what comes next makes ex-communication look like a mild dressing down.'

***

Even after ten days of getting 'sorted out' and making several sorties down what, for obvious reasons, westerners called Sewer Street to stock up for the Yanbu trip, my little faux pas was still the talk of the crew.

'Can't believe you tried that,' said Mick the mild-mannered mechanic from Middlesborough who'd volunteered to help us get set up at the coast. 'Respect. But if you'd be so kind, please don't go trying it on again while we're on the same crew. I've got just weeks to go before I'm out of here and the wife's not going to be happy if I'm detained on suspicion of helping someone else break Islamic law. Not happy at all.'

Got it, I gestured back. But it was going to be difficult, especially resisting the come hither eyes of the women in the souk, the only bit of them not veiled by the ubiquitous black niqab.

'Forget it,' cautioned Mick. 'For my sake at least. Wife wants all of me back. And not in two pieces.'

I was still trying to forget it as we loaded the necessary survey gear into three beaten up Toyota pickups for the two-day trip along the undeviatingly straight thousand kilometre-long desert highway to Jeddah. Those eyes had been so beguiling it left me wondering why, if Islamic law required women to cover themselves completely for the sake of modesty, that didn't apply to eyes as well. It didn't make any sense.

'Not a lot does here,' shrugged Mick as he helped me load my gear into his pickup to the envy of all others on the crew. After evading an Islamic lynching by the skin of my teeth, it'd been decided I must have the luck of Old Nick and every one of this superstitious lot wanted such a talisman as a passenger in their own vehicle. His presence might help prevent coming into overly close contact with the roadfuls of mad Bedouin families barrelling along in suspensionless pickups with camels in the back and zero experience of highway driving in the front.

As things turned out the luck I'd apparently brought with me had, if you believed the crew chatter, been partially responsible for making it to Yanbu without serious highway incident. Only partially because at one stage it was only Mick's rally driving skills that prevented us becoming the cause of a serious international incident, crashing through the gates of a police post set up to stop infidels entering the Islamic holy of holies, Mecca. My presence in the lead car had done nothing to prevent that and as such maybe I wasn't quite the talismanic force they'd thought me to be.

After a day and a half of trying to stay awake on the uneeringly bend-free road west from Riyadh to Jeddah and a night bivouacing beside it, it'd come as a massive relief to reach the magnificent Taif escarpment down to the coast off the central desert plain. At least the terror of negotiating the continual and never signposted hairpin bends down the escarpment would keep you from dropping off, something that was guaranteed if you did drop off.

Unfortunately it also meant that, in having to keep one's eyes fixed on the road, they were distracted from seeing the one warning sign that did exist and HAD to be noticed. The one reading 'Mecca – Non-Muslims Prohibited Beyond This Point'.

Careering down the slope, without any prior warning we were on it. A green-painted gate at the foot of the escarpment fronted by a right angle bend that traffic containing infidels like us had to negotiate if they wanted to avoid having to explain their surprise arrival to the guardians of Islamic purity manning the police post. Police who were taking bets, I wouldn't wonder, on whether we were going to make the turn without mishap.

That we did must have come as much of a surprise to them as it did to us, especially driver Mick who, in an impressive piece of driving dexterity, somehow managed to slew round it on two wheels to bring the vehicle to a shuddering halt some distance on.

Bouncing back down onto four wheels, he gradually brought the vehicle back under control, pulled in and remained rigidly pinned to his seat, eyes staring maniacally through the windscreen, white knuckles gripped quiveringly round the steering wheel.

'Know that hooch you tried to smuggle in?' he finally managed through lips frozen to his teeth in the rictus grin of a trainee astronaut undergoing his first flight in a G-force simulating centrifuge. 'Got any left?'

Chapter 4.2

By the time we'd reached Jeddah, in those days just a hot dry dusty port city with crumbling building after crumbling building flanking a ramshackle assortment of decrepit wharves, it was clear the crew had begun revising their opinion of me. After the Mecca incident I might be as much trouble magnet as talismanic force and as such maybe best for me to stay in Mick's cab as the convoy negotiated its way through the mad Jeddah traffic to the tumbledown no stars 'hotel' Genghis Carne had ordered us to doss down in for the night.

By the time we found it, it wouldn't have mattered if it'd been a snake-infested garden shed, so exhausted were we after two full days on the road under a desert sun. So after a quick bite to eat in a local restaurant, by nine o'clock we were all snoring so soundly in our flea-ridden pits that even a tsunami warning wouldn't have roused us.

Did they have tsunamis in the Red Sea, I wondered aloud as we set off on the four hundred kilometre trip up a blasted coastal highway to our final destination the next morning? If they did, that might affect our work schedule a bit, the site being located within spitting distance of the sea.

No one seemed to know which was all a bit worrying. Given my record of attracting trouble, if this coast had ever been inundated a repeat performance while we were in residence was almost inevitable leaving Genghis with the solemn task of informing our next of kin of the entire crew's disappearance without trace.

With that in mind it seemed sensible to choose an elevated spot for the survey camp, my responsibility on arriving first at the site.

Good theory, shame about the practice. For after an hour of criss-crossing the bone dry gravely site and failing to find anything even resembling a small hillock we gave up. The whole place was as flat as a hedgehog roadkill, nowhere higher than a couple of metres above sea level, and in the end we just stopped, offloaded the tents and awaited the inevitable gripes over the choice.

Mick and I knew they'd be coming. It's what surveyors do, get picky about things. Something to do with having to be so precise when it comes to closing out a triangulation to within tiny allowable tolerances, I think it is. So in effect it didn't really matter where I or anyone else chose as the campsite. Someone would always find fault and the same, I knew, would apply to the choice of survey technique best suited to the job.

I wasn't wrong. Gathered around the communal table in the army surplus mess tent over dry rations and warm Fanta that evening just one agreement emerged – that, in the event of there being no clear agreement on which survey technique to go with, the one with the most extensive experience of desert conditions should have the casting vote.

Phew. Thankfully that counted me out. After already having had to field the mutterings over the campsite choice I had no interest in having them added to with grizzles about my preferred choice of survey technique. Former Royal Engineers desert rat second lieutenant 'Jonno' Groats could have that pleasure and he was welcome to it.

***

Also known as BTB by those who'd worked with him before – courtesy of his predilection for never doing anything that wasn't By The Book – our newly-elected team leader lost no time in formulating a plan no one else agreed with.

You could have heard the groans in Riyadh when Groats announced his decision – a simple geomorphological survey identifying and recording the position of every pimple in the landscape from the fat file of aerial photographs that'd come with the survey contract. Or, in other words, one of the most tediously repetitive survey techniques ever invented.

Bar Groats, there wasn't one amongst us who didn't dread having to be so pedantically precise in the crossing of t's and dotting of i's on this job. Ranked solidly bottom on everyone's scale of survey job enjoyment except Groats's, once started any survey using this method was well-known for rapidly descending into months of just going mindlessly through the motions, a degree of tedium that in this case even the knowledge of knowing we were in possession of highly classified documentation couldn't break.

In Saudi Arabia, we were fully aware, all aerial photographs were the strict property of the country's defence establishment and being given access to them instantly made us a potential target for any alien force desperate to grab a worthless slice of Arabian desert… thus, in effect, making us jointly and severally responsible for the defence of the Saudi realm.

Charged with that onerous responsibility, it was a wonder any of us managed to sleep at night but somehow we did. Primarily because it'd been unanimously decided behind Groats's back that should any insurgent creep up on us in the desert darkness with the intention of capturing our precious photographic collection they could have the damn things. Without them, perhaps we could get back to doing some proper survey work.

***

Or we could providing we could get the vehicles running. Being forced to use Toyota pickups in place of the surveyor's standard Landrover workhorse – banned in Saudi because of some obscure Israel/Landrover connection – we knew we were getting seriously substandard modes of transport.

Just how substandard only became apparent the day Mick the mechanic left. Having assured us the two Toyotas he was leaving us with had received enough tender loving care to see us through to the end of the job, he drove off in the third – back to Riyadh and thence to his beloved Middlesborough – and almost instantly the first of our two rebelled, resolutely refusing to start.

Only through our own amateurish mechanical efforts was it coaxed back to life but it set the tone for the remainder of the job. If it wasn't one needing our attention it was the other and as the survey work downtime days turned to weeks we began to fret. At the current rate, we calculated, the number of overrun contingency days allowed for this job would be exhausted before we were even close to finishing and even BTB had started admitting something had to be done.

'Someone's got to let Genghis know,' said Salt and Pepper simultaneously, the team's pair of Mr Potatohead lookalikes who seemed to have developed a sort of telepathic understanding during their extensive time in Saudi and had thus been dubbed The Condiments.

'Agreed,' said Groats. 'Who's it going to be?'

'The team leader?' I ventured. 'Can't be me. Got that carburettor to sort out and The Condiments went last time. Anyway, best coming from the big cheese, wouldn't you say? Carries more weight.'

You could almost see BTB's chest swell on being described as one who carried 'weight'. It made up for his garden gnome stature and inability to grow a decent moustache.

'Gotcha,' the looks that passed between the rest of us said as Groats fell hook line and sinker for the compliments ruse, almost readily agreeing he was the one for the job. 'Never learns does he?' the looks said.

Had the trip to what was loosely known as the nearest 'town' been a thing of pleasure the ruse employed would have been somewhat different, one that'd have left our dear leader convinced he was so indispensable to the survey task he couldn't possibly be spared from the field.

But faced with twenty kilometres of being choked to death by the desert dust to reach the main road then a game of Russian roulette with the lunatic Bedouin-dominated highway traffic for a further fifty clicks to get to a settlement whose only attribute was a telephone that worked one day in three, it was not a trip that got many volunteers. Even staying in a camp running perilously short of food, water and Fanta was preferable to that. At least here we had the sea to cool off in together with a beach, one that stretched all the way to Dhahran on the other side of the Arabian peninsula.

Had we had a radio, of course, no trips to phone progress reports in to base would have been necessary. But in its absence – a restriction forced on us by the government's point blank refusal to let us have a transmitter radio on the grounds of national security – there was no alternative. From time to time one of us had to make the trip to phone Genghis, a daunting enough prospect even when there was no bad news to report.

Not the most affable of human beings and the living proof of the old adage that, when it came to one's superiors, ears have walls, after having to visit the 'town's' public phone office every hour on the hour for days on end to get a connection it wasn't unusual to finally get through to find one's own ears being chewed off for not reporting in sooner.

So, with the job of having to report that this job was likely to overrun being about as popular as Genghis himself, this particular trip had produced even fewer volunteers than usual and it was with some relief that we waved BTB and our resident cook/interpreter Abdul the Egyptian on their way, silently congratulating ourselves for our perspicacity in having elected him to the post of Surveymeister-General.

Five days later we were feeling rather less pleased with ourselves. The pair still weren't back and we'd starting getting concerned. The camp stocks of tinned sardines and Fanta were running low and unless BTB and Abdul appeared over the horizon soon with new supplies we'd be having to resort to the tins of dubious vintage corned beef and bottles of eerily luminous green liquid with unreadable Chinese labels that had lain untouched and forgotten in a corner of the oven-temperature foodstore tent ever since our arrival four months earlier.

Fuel was running low too, the level of diesel in the camp's forty-five gallon oil drum down to the point of risking bringing up sediment guaranteed to clog the now solitary Toyota's fuel lines.

That hadn't been a major concern for the first three days of being left in glorious isolation. Three days in and out from town was the norm, two a record, one unheard of and four leaving the others suspecting the trip-maker must have discovered some attraction in town they'd missed.

So when day five came and went with no sign of our supplies appearing we found ourselves in uncharted waters, familiar ground you'd have thought for the exploration surveyor but on this occasion ground that saw anxiety levels begin to rise to the point of entering further unknown territory, the need to make a decision.

Sipping at straws in three of the last remaining bottles of Fanta that evening, it was, we decided, time to stop pretending to work and just do it. Stop, that is. What fuel we had left needed to be conserved for our escape should BTB never reappear.

Even as we said it, the irony of the decision brought a collective smile to our faces. In a country awash with hydrocarbons we must be the only people in it having to conserve the stuff, a thought that left us staring silently into our bottles shaking our sun-scorched incredulous heads.

It was Pepper who broke the silence. Surely, he said with

a slightly strange look in his eyes, with Saudi Arabia all but swimming in oil all we had to do was dig a hole! Just a few metres deep should be enough to make a strike and then we'd be saved. SAVED!

Salt and I just looked at one another. We'd been wondering who would crack first. Well now we knew and that meant just one thing. Pepper would now have to be handled with extreme care.

'Don't do anything that might send him over the top into funny farm territory,' Salt mouthed at me as Pepper's attention was taken staring wild-eyed at the last few bottles of Fanta in the fridge. 'Don't want him getting any thoughts of cannibalism!'

Shit, I thought. Knowing how the pair had reached the stage of virtual telepathic communication, of one thing I could be very sure. I'd be the last to know if that was exactly what was going through their respective but connected minds.

Chapter 4.3

Did I have any real cause to worry about The Condiments' cannibalistic tendencies? Fortunately, I was never to find out. BTB's arrival on day seven saved me from having to sleep with one eye open and as he pulled into camp accompanied by another, almost gleaming, vehicle I all but hugged him.

'And on the seventh day...' I began, '...God discovered oil!' beamed Pepper, arms wrapped lovingly around the new barrel of diesel on the back of Groats's pickup.

The only one not getting the joke was Genghis, a look of undisguised disgust shrouding his face as he alighted from the virtually new Nissan he was driving.

'So this is what you call a professional set-up, is it?' he growled, casting a jaundiced eye around the squatter camp we'd called home for the past few months. 'If I didn't know pigs were banned in Saudi...'

'... you'd be the first to ask for a bacon sandwich,' I thought as Abdul and I meekly followed instructions and moved the tent Genghis had selected to be his personal accommodation to a spot some distance from the rest of us. 'That'd about sum up everything you know about Saudi and the conditions your precious bloody company expects its workers to put up with.'

It was an unfair criticism, of course. If there was anyone who DID know everything there was to know about this country and how to use it to his best personal advantage it was Genghis Carne... something that was to become all too self-evident in the hours between his arrival in camp and a hasty departure made necessary by having the real reason for his visit rumbled.

Looking back I realised alarm bells should have started ringing the moment I saw Genghis wasn't really listening to Groats's horror story of getting to Yanbu to make his report to base. If he'd really made the trip to see for himself whether Groats's report about the condition of the vehicles we'd been lumbered with needed acting on, he'd have been as gripped as the rest of us as Groats ran through the sequence of events that'd led to having to abandon his vehicle by the side of the road and hitch a ride with a passing Bedouin the rest of the way to town.

But he wasn't and now, thanks to a chance sighting of a document in Genghis's possession, I knew why. He wasn't interested in us or our vehicles or in the conditions we had to work in. All he'd come to do was assess progress on a job that was, in fact, about as far from one with no oil industry connection as it was possible to get.

***

As soon as Genghis had glided into Yanbu in the air-conditioned comfort of the spanking new four-wheel drive he'd hired after flying down to Jeddah, the extent of our plight should have become blindingly obvious to him. Groats's first request on meeting him was for a lift to get back to recover his abandoned Toyota and get it to a local mechanic.

But according to Groats the man had been about as concerned by our circumstances then as he was in the mess tent later, reacting to Groats's story by doing no more than fix his dispassionate gaze on the untouched plate of the usual unrecognisable mishmash of offerings from Abdul's kitchen that he'd pushed contemptuously away.

With his mind clearly elsewhere, Genghis's face remained stoically impassive as Groats recounted how he'd finally breathed out on reaching the highway. With the Toyota starting to cough and misfire in the desert dust en route to the main road, Groats said he'd thought the worst was probably now over. The dust in the vehicle's tubes would gradually dissipate and pretty soon the damn thing would stop kangarooing along the road.

He was right. In the end it did and Groats had managed to turn his attention to simply surviving the devil's own highway the rest of the way to Yanbu.

All had apparently gone well for a couple of kilometres, the vehicle behaving itself and the traffic showing less sign of suicidal tendencies than usual. But then there'd been a loud bang and all Groats's attention had been diverted to keeping the vehicle from slewing into oncoming traffic.

It'd been a blow-out, of course, something that was easily dealt with and something that almost made Groats relax. If this was the worst that was going to happen, the rest of the trip should be a breeze.

It was as if the pickup had been reading his mind, said Groats. For once he and Abdul had got the machine jacked up to replace the flat with one of two spares being carried, a tyre on the opposite side burst.

OK, Groats had told Abdul as they replaced that one as well, without any more spares they'd have to take it very easy the rest of the way and Groats had climbed back aboard to start the engine.

Nothing. No ignition. Not a peep. Bugger. Now what?

Abdul knew. Flag down one of the mad Bedouins streaking down the road with more concern for the welfare of their camels in the back than for others on the road.

With some reluctance Groats had agreed and within minutes they were crammed in with said camels for the remainder of the trip to Yanbu. There he'd found an available mechanic but one whose own pickup was off the road and couldn't attend Groats's vehicle until he'd fixed his own.

How long was that expected to take, he'd enquired via Abdul?

Two days, Abdul had translated back.

'Inshallah,' the mechanic had then added.

'Inshallah,' Abdul had translated helpfully.

Well, thought Groats, since it'd probably take that long to get through to Riyadh on the phone anyway then OK, he'd roll with it but that left the pair in need of a place to stay.

'Why not here?' asked Abdul after the mechanic had offered a room. 'This man says we won't find better.'

Too exhausted to argue, Groats had wearily nodded his agreement and the pair were soon installed in 'the best room in the house', one that looked like its previous occupant had been a goat. Or even a groat, I thought uncharitably as the story progressed.

To Groats's astonishment he'd got a phone connection to base in moments the next day and had secured Genghis's agreement to come to see what could be done. He'd be able to fly down in a couple of days so Groats should remain where he was and await his arrival.

Three days later Genghis had swanned into town, gawped at where Groats and Abdul were staying, picked up the mechanic whose vehicle the Will of Allah had decreed hadn't yet been fixed and the whole complement had relocated back to Groats's vehicle, found to be fully intact save for anything portable that'd had to be left behind including both flat tyres.

'Well, at least they left you the wheels on the hubs,' Genghis had observed as the mechanic fiddled with the Toyota's electrics.

After an hour of fiddling, the engine finally cooperated and both vehicles edged their way gingerly back to town to go in search of some new spare tyres.

'No problem,' Abdul had translated after a brief conversation with the mechanic. 'This effendi thinks he has some in his parts store.'

Indeed he had. A pair that looked suspiciously like the ones that'd gone missing from Groats's vehicle, both now repaired, inflated and ready for use.

Unable to prove anything and, with the mechanic just smiling an enigmatic smile on being informed by Groats that these wheels looked somewhat familiar, a deal had reluctantly been struck, the spare tyres loaded back on the Toyota and Groats and Genghis had set off in search of supplies for the camp leaving the mechanic smirking as he pocketed the roll of Saudi riyals Groats had had no option but to hand over.

Almost to Groats's annoyance, he confided to me later, his Toyota had behaved impeccably all the way back to camp, even the tyres Groats had thought would succumb under the load of the camp supplies and a new barrel of diesel. If at least one of them had had the decency to burst it'd have helped his case in persuading Genghis to organise some replacement vehicles. But none had and that left Groats hoping one or other of the camp Toyotas would cooperate and break down while Genghis was on site.

Subsequent events were to leave both us and Genghis disappointed, but for different reasons. As if the vehicles had jointly decided to spite us, neither showed any sign of mechanical malfunction during our boss's two-days of inspecting every corner of the site and in the end he'd stormed off fuming over having been begged to come all this way for no good reason.

At first Groats just watched him go, eyes glazing over as the full implications of Genghis's parting, stinging, remark sank in. In his considered opinion, our visitor had barked on gunning the Nissan's engine and roaring off up the track, the only thing here that needed replacing was the man who'd duped him into embarking on such a fool's errand.

For a while Groats just stood there shrouded in Genghis's dust, upper lip stiffening in the manner required of him in his army days.

Then, to our collective amazement, he turned and treated us to one of the most outspoken outbursts any of us had ever witnessed from anyone attached to the company.

He had, he said with fists clenched and body visibly quivering with rage, done his very best under trying circumstances. He'd tried to be a good team leader, he really had, doing everything by the book and reporting in faithfully and without embellishment his honest assessment of the state of affairs apropos this project. But all his efforts obviously hadn't been enough and with that in mind the time had clearly come for him to reassess his position.

Oh dear God, the rest of us thought. He's going to ask one of us to take his place, the last thing any of us wants. With Groats self-demoted to mere foot soldier it'd be one of us in Genghis's line of fire instead.

As we waited for the denouement, every one of us holding our breath and hoping to hell he'd be choosing someone else as his heir apparent, in our joint perplexed state we almost missed the sting in the tail Groats was holding back for last.

'So,' he said finally, 'if what I've just heard is an accurate measure of how this company regards those in its employ who're doing their best to do the company's bidding, I see no option but to give them what they want and vacate my position at the earliest opportunity.'

'Earliest opportunity?' said Salt, confused.

'The moment my contractual term on this project expires.'

What was he saying, the looks that passed between the rest of us said? That he'd be quitting as team leader but not until the end of the project?

'Yes,' he said when I asked, hoping to hell that's exactly what he meant. 'I'll be submitting my resignation as team leader at the end of the project... or at the termination of my contractual term working on it, whichever is the sooner.'

'But that's just a few weeks away,' said Salt. 'All our contractual terms are. What if the project isn't finished by then?'

'Then tough shit,' said Groats flatly in the most militant tone I'd ever heard issuing from his mouth. 'They can find themselves another patsy to finish it. After what Genghis just said, once my contractual six months is up I'm done with this damn project… and with this company if they get shitty about me not agreeing to extend or renew. There are other survey companies out there, every one of them crying out for experienced personnel.'

Jayzus, I thought. He really is pissed off. To the point of risking getting summarily suitcased if OTC took exception to being dictated to. If that was the eventual outcome, I warned, he could well be risking ending up with such a black mark on his record others might be reluctant to take him on.

'Correct on all counts,' said Groats, 'which is why I'll be working out my six-month contractual term. Once that's done with they wouldn't have a legal leg to stand on for firing me. Having fulfilled all the terms under my contract, if I decide not to either extend or renew it that'd mean I'd just resigned my position, not been sacked.'

Holy crap, I thought as Groats effectively announced without knowing it that he was joining a resignation team of which I'd thought I was the only member. Now it wouldn't be just me asking for my P45 back at the end of this job. Now there were two of us but for two polar opposite reasons.

Groats was going because he could no longer work for a company that seemed to no longer trust him. My reason for quitting was also on a point of trust but one in which it was my trust in the company that'd disintegrated, not the other way round. Or more accurately, that I did still trust OTC but only to do one thing – to lie and lie and lie again to those in its employ through its dirty avaricious teeth. After what I'd seen poking out of Genghis's briefcase on passing by his unoccupied tent en route to relieve myself during dinner the previous evening, of that there could now be not an iota of doubt.

Chapter 4.4

'Dumbass!' the document's spine screamed at me from its not very well concealed hiding place in Genghis's tent. 'I CANNOT believe you fell for all that horseshit about being put on a non-oil project. An industrial complex in Saudi Arabia WITHOUT an oil industry connection? That'd be like having a pub without a brewery. Well, maybe not in Saudi Arabia but you know what I mean. It just ain't gonna happen pal and I'm amazed you didn't spot it the moment you were told where it was you'd be working!'

'So what is it then, this industrial complex?'

'JEEEZUS! A petrochemical plant, of course, dumbass! What else COULD it be in Saudi bloody Arabia? Read my spine! What does it say?'

"Yanbu Petrochemical Complex – Schematic Draft Version 1.0."

'And what else does it say?'

'Nothing. Those are the only words.'

'Crap. Read between the words and I'll tell you what it says. It says you're working for a company that can be trusted about as far as a backstreet Saudi mechanic who's just discovered there's a foreigner's vehicle left lying unattended by the side of a remote stretch of unlit desert road. THAT'S what it says, dumbass? That you've been played, you pitiful prevaricating pillock, and if you want my opinion it's time you stopped fucking about and actually did something about it.'

Dragging my stunned heels back across the gravelly stretch

between Genghis's tent and the mess tent it took time for what I'd discovered to sink in. Not again! Surely! This time I'd been so sure. This time the company had been straight with me. I'd have put money on it. But now the truth was out. OTC simply could not be trusted and the document was right. It really was time to stop fucking about. Time to properly bite the bullet I'd only nibbled at in Gabon and tell this conniving bunch of double-dealing carpetbaggers where they could stick their bloody job.

But not just yet. Just one thing would result from giving company lickspittle Genghis both barrels at the dinner table – a swift departure back to join the Brighton dole queue. Letting on I'd been rifling through the contents of his briefcase – a thing I'd wondered why he'd not only brought but kept permanently close by him as he followed us around in his gleaming Nissan on a tour of the entire site during the day – was guaranteed to get me handed my own baggage. My suitcase.

So no, I thought. First I'd play the Gabon card. Get to the end of my contractual term then stick two fingers up at them. As Groats had realised and was to echo the following day, it was the only way of avoiding finding oneself blackballed by the entire survey industry.

Even so, somehow Genghis had to be left in no doubt that his and the company's little wheeze had been rumbled. After being assured I'd no longer be being put on oil industry-related jobs and then finding that's exactly what I'd been put on I owed it to myself big time. But how?

It was Genghis himself who handed me the opportunity. Loathe to talk about anything but work, he'd proceeded to dominate the dinner table banter with inane drivel about how we'd all be winning feathers in our caps if we brought the job home ahead of schedule. The client was itching to get on with the construction phase and needed our maps soonest to get everything in the pipeline flowing.

'Pipeline?' I echoed with one eyebrow raised.

'Figure of speech,' said Genghis blanching slightly. 'Shorthand for getting cracking.'

'What, as in hydrocarbon processing? "Cracking" makes it sound like the survey we're working on is for some kind of oil refinery. But that can't be right. I was told it had nothing to do with oil, this project. Has something changed?'

'No change that I've been told about,' he mumbled into the bottle in front of him, looking anywhere except into my eyes. 'Still the original scheme so far as I'm aware although what that is I've no idea. Way above my pay grade that sort of info,' he went on, lying comprehensively through his teeth. 'As just a company functionary all I'm charged with is getting the survey done pronto.

'So as of tomorrow,' he added quickly having seen a way out of his faux pas, 'I want to see things speeded up. And just to give you a bit of motivation, if you beat the deadline there'll be some 'iced tea' waiting for you at the Wimpey bar. That should focus minds a bit.' And with that he rose from the table, mumbled something about needing to pack as he was off early in the morning and scurried off to his tent, avoiding eye contact with anyone as he went.

'Well,' said BTB after he'd gone. 'You heard the man. If that 'iced tea' is to be ours we've got to get... er... cracking. So maybe an early night? After these past few days I know I could do with one.'

Personally, I wasn't as keen as BTB on turning in early. I knew what awaited. A night of tossing and turning as my mind churned with all the new developments of the day. Although Genghis hadn't exactly confirmed the survey was oil industry-related, neither had he denied it and the shifty evasive bastard had made himself scarce before I'd had the chance to drop any more hints about him knowing exactly what the survey was for. No doubt he'd now be zipped firmly up in his tent making sure the file I'd seen was locked securely away in his briefcase. After being cornered into having to state publicly that he had no idea what the purpose of the survey was, if any of the mere mortals who were probably best avoided for his remaining hours on site saw that file it might complicate matters somewhat.

***

With or without the motivational offer of 'iced tea' waiting for us at the Wimpey bar – the code Riyadh expats used for the illicit brews concocted behind closed doors at every one of the city's western company compounds – it'd have been a miracle if we'd got anywhere close to finishing the project in the few weeks remaining before the team's contractual terms expired. Not unexpectedly, the moment Genghis was out of sight both Toyotas returned to their usual levels of reliability and if anything the pace of survey progress in subsequent weeks slowed even further.

Genghis, of course, blamed it on BTB's lack of organisational and motivational skills, BTB blamed it on the state of the vehicles and the rest of us just stood around looking up to the heavens. Our Moslem hosts were right, we'd decided one night after a particularly harrowing day of breakdown after breakdown producing a daily progress rate of precisely zero. This job would only get done if and when Allah willed it and if that meant the task of tying up the loose ends fell to someone else in the absence of the contract-expired original crew members, then who were we to take issue with something the great Islamic guiding spirit in the sky had decreed?

Needless to say, our superiors didn't see it that way. Particularly when BTB reported in that, owing to the fact that our contracts were due to expire in a few days time, we'd be returning to base forthwith to begin the process of booking our flights home.

You could almost hear the fuses popping in Genghis's brain on receiving Groats's final report from the field, one which every one of us had made the trip to Yanbu to listen in on.

'But... but... but... you're not finished yet!' he yelled down the line. 'Your job is to stay with the project until you are!'

'Not according to the wording of the contract,' Groats had explained calmly and patiently to a man who's temples we could almost see throbbing ever more violently the more measured Groats's tone became. 'There's nothing in it about having to stay until completion. Just that we're contracted to work on the project for six months. We've done that and now we're going home.'

'But... but... company LOYALTY,' Genghis spluttered back. 'Does that mean NOTHING to you? Don't you think you OWE it to your employers to see things through to completion? They've looked after you. Surely you can see that that's a favour worth returning!'

'All that has been taken into account in our deliberations,' replied BTB firmly but equably, 'and so far as everyone here is concerned we're fully prepared to be as loyal to the company as the company has been to us. Which is why, as of tomorrow, we're packing up and coming in... vehicle condition permitting.'

For a while the line went silent as Genghis, we knew, wrestled with himself over whether and how to resort to the 'p' word. Not coming naturally to a man to whom p's & q's were not a standard part of his managerial technique, in the end he gave up trying to get his mouth around the word 'please' and opted instead for the only firearm in his armoury – all out aggression.

'NEGATIVE!' he shouted down the line. 'You'll bloody stay where you are and if you don't I'll have you all keelhauled, you treacherous bunch of bilge rats. Show up here and...'

We never did get to hear what he had planned for us on our arrival. With a small smile, Groats had gently replaced the receiver while Genghis was in full rant, turning to us as he did so and mouthing that that was quite enough of that. We had preparations to make.

***

Like horses knowing they were heading back to the stable and a nice bag of hay, both Toyotas were on their very best behaviour all the way back to Riyadh, even negotiating the perpetual hairpins up the Taif escarpment without complaining to reach the roadhouse at the top for a much-needed overnight stop.

Once installed, we gathered round a table to plan our game strategy for the next day. Over some quite palatable fiery Arab fare Groats first outlined his own position then, in true democratic fashion in a country where such things were as foreign to its residents as decent bread and cheese, he invited us to state ours.

'For my part,' he said, 'after a visit to the travel agent to check on flights to London I intend to visit Genghis's office for only as long as it takes to formally re-state everything I've already said on the phone. Once he's blown his final fuse, I'll then be heading straight to the airport. Under the terms of the contract he'll have no grounds for either summarily dismissing me or trying to prevent my departure and if he opts for either, my response will be straightforward. I'll simply down tools and refuse to work until he agrees to allow my repatriation unhindered. I don't think it'll take too long for him to cave in. OTC doesn't take kindly to their managers keeping freeloaders on the payroll, taking up space in the bunkhouse and receiving sustenance at the company's expense.

'Once home, I'll be doing no more than wait for the company to contact me. If they're prepared to see sense and demonstrate that they understand why I decided to play by the letter of the contract and exit the country at its termination then all well and good. Any further offers of employment on other projects will be considered on their merits.

'Failing that, I'll be starting an immediate search for an alternative employer… particularly if OTC even hints at dismissing me. Under UK employment and contract law, they wouldn't have a leg to stand on for taking that action and if they try it I'll be instantly putting the matter in the hands of m'learned friend.

'So that's my strategy. The floor is now open to anyone with anything further to say.'

Personally, I didn't. Groats and I had discussed it at length while sharing the driving from Yanbu to Jeddah, round Mecca and up the escarpment to the relatively cool conditions away from the steaming Red Sea coast and after some thought I'd decided there wasn't a lot in his strategy with which I could find fault. He'd effectively been echoing everything I was planning to do with the possible exception of joining a union which, I supposed, I could turn to for help bringing a case for unfair dismissal if that's the stunt OTC tried to pull.

The Condiments, on the other hand, did have something to add. After consultations with Abdul, crammed into the cab of their vehicle in the absence of the third vehicle we'd travelled down with, their strategy, they'd decided, would be slightly different. With all three of them sharing a common desire to remain in Saudi for financial reasons, they'd decided that a more conciliatory approach was needed. One in which they planned to offer to enter into negotiations with the company over returning to site.

It all hinged, they said, on the company agreeing to provide better vehicles and to assign a trained mechanic to the job to keep them serviceable. Then they'd be fully prepared to return to Yanbu to see the project through. They had nothing against Genghis or the company per se and were quite willing to demonstrate their loyalty providing their conditions were met.

Stifling a burning desire to mutter 'Judas' as they laid their thoughts out for our consideration, I wondered grimly how that would affect Groats and me. We'd been banking on a show of crew solidarity in the face of company intransigence and this would hardly help our cause. Genghis could use it to undermine our case against working on until the project was done with. 'See?' I could visualise him telling us. 'If they can agree to extend a bit to see things through, why not you? Some sort of commie agitators are you?'

'Not until I met you,' I could hear myself responding in my mind's eye. 'I was no militant until you came into my life. But now, thanks to you and this shoddy little outfit, I've seen the light. Credit where it's due. Without the enlightenment gleaned from my time with this company I might be still labouring under the illusion that every commercial organisation is basically honourable, always having its employees interests at heart. So thanks for that. After everything I've learned from you now I know better. Now I know some companies feel they have the God-given right to treat their workers like slave-labour mushrooms, keeping them in the dark and feeding them on bullshit.

'But it's not all one way traffic, Mr 'Genghis' Carne. While it's definitely been a learning experience for me, perhaps it has been for you too. Dare one hope that someone somewhere might have learned that treating one's employees like shit can result in that shit being thrown back in their faces. If they have, then maybe they might also have learned that a bride can only take so much of being left at the altar. Do it too many times and you shouldn't be surprised if one day that favour is returned in spades.

'Well that day has now come, motherfucker. As of now, this bride is walking back up the aisle. You can go find yourself another patsy to do your dirty work, you slimy conniving piece of shit. Treat THEM like a cunt. This one's been fucked over once too often.'

Chapter 4.5

'Good God! You really said that to his face?' asked the intense angular woman who'd cornered me at a party a few weeks after getting home and had shown interest in hearing about my travels. 'Respect.'

'Sadly not, I'm afraid. The man didn't give me the chance. Nor Groats. After marching into his office fully prepared for a stand-up slanging match, all Genghis did was hand us our tickets and passports, muttered something obscene under his breath and waved us out of his office. No comment. No argument. Nothing. Something told us head office had ordered him not to get into any further discussion over our actions.'

'So what happened next?'

'Got a taxi to the airport, flew home and went to the pub after reading the letter from the company neither I nor Groats was surprised to find waiting for us. Letters of dismissal they were although they'd sneakily tried to cover that up with some pretty artless wording. Due to 'difficulties arising out of your indifference towards the companies activities,' mine read, 'the company finds itself unable to offer you any further contract work.''

'Sounds like dismissal to me,' said Ms Angularity.

'Me too, so I took it up with the union I'd joined the moment I got that letter. It took no more than a cursory look for them to decide that's exactly what it looked like and they expressed an interest in taking up my case. Definitely a blatant example of unfair dismissal, they concluded after I'd filled them in on the details.'

'So there'll be a hearing of some sort?'

'Nope. No court case either or even an out-of-court settlement. After further consideration the union lawyer decided we didn't have a case after all. Or rather we did but that no representation could be made on the grounds that everything that had happened had been outside the country. Apparently, UK legislation covering employment rights doesn't cover people working outside the UK, even if they're British citizens working for a British-registered company.'

'You're kidding!'

'Nope. Overseas workers just aren't covered and as a result there's no recourse to the courts or even the arbitration panel. They can be hired and fired at will and the companies concerned can get away with it scot-free.'

'Good Lord! Didn't know that. What're you going to do about it?'

'Nothing I can do. And the scumbags knew that all the time. Which is why they didn't even try to talk me and Groats round. Their attitude is that if someone's pissing into your tent and it's clear no amount of persuasion will get them to come inside and piss out, piss seriously back with a big corporate dick. That should dampen their ardour enough to persuade them to go elsewhere to relieve themselves.'

'That'd be enough to piss anyone off...'

'Enough to wet my pants some days. But since there's nothing I can do about it, I just have to ride it out. Anyway, I'm glad to be out of it. I've done my time with that industry. It's time to look elsewhere for gainful employment.

'It's Jonno Groats I feel most sorry for. If he's unable to bring a case for unfair dismissal he's fucked. So far as I know he doesn't have a Plan B. He's a surveyor through and through and wants to stay in the business. But it's unlikely any reputable survey company will have the poor bastard with this on his record. Maybe the likes of 3S. They'd take anyone. But no one regarded as a proper professional operation.'

'So what's your Plan B then? Not being intrusive or anything but never met an oil explorer before. Sounds such a specialised profession I doubt there'd be many employment options outside what you've already been doing.'

'Not sure at the moment. Got enough funds stashed away to give me time to think about it but nothing immediately springs to mind... except that I'm definitely done with anything even loosely connected to the oil industry. Gotta be something out there where questioning the status quo wins you kudos rather than the sack. It just hasn't revealed itself to me yet.'

***

'What about journalism?' she said after we'd finished exploring one another's brains out back at her flat. 'Questioning things is what journalists do.'

'Were you thinking about that all the time we were…?'

'NO! Well, not all the time. It just occurred to me that you might have an aptitude for it. And if you did I might be able to help. Got some friends in the business I could introduce you to if you like. Wouldn't hurt to have a chat, would it?'

'Not sure what we'd chat about. Don't know anything about journalism.'

'Don't have to. They'll probably do most of the talking. Telling others all about themselves is a journalist's favourite pastime.'

'Hmmm. Not sure. I'd have to think about it.'

'Oh, come on!' she wheedled, moving to mount yet another explorative assault on my person with the intention of reaching climactic contentment at the expense of the near-wasted mere mortal wilting under the effects of her concerted and irresistible pincer movement. 'You know you want to!'

'All right! All right! You win. I give in!' I whimpered as Ms Angularity's ground offensive reached its peak. Anything to win a reprieve from descending even further towards a demise I'd once dreamt would be the perfect way to go. 'Set up your damn meeting. I'll do anything you ask. Just let me get some sleep!'

***

And so it was, Sir Isaac, that my metamorphosis from grubby oilfield worker to grubby Fleet Street hack began. A metamorphosis that once you've heard the details, I'd humbly suggest confirms my full adherence to the action/reaction premise contained in your Third Law of Motion. As you'll see, my reaction to realising I was integral to the oil and survey worlds' planet-wrecking actions was to initiate a handbrake turn, travelling through a full one hundred and eighty degrees to make amends for my own actions through helping in my own small way the campaign to put a stop to such wanton, thoughtless destruction.

The road towards that end started with my agreeing to write – with Ms Angularity's intense, not uncritical, assistance – a feature on the life of the average oil exploration worker for a magazine I'd never heard of but which, she assured me, would be a stepping stone to bigger things.

She was right. Bigger things did come, and with one bigger thing leading to another, a position with one of the UK's quality nationals was eventually offered, the job title of which brought an ironic smile to my lips on being informed of the area they wanted me to cover. After once being pilloried for daring to question the environmental credentials of the likes of Shell, as one of the UK's first-ever reporters assigned the job of covering issues associated with energy and the environment now someone would be paying me to do precisely that.

Forty years on and the environment still dominates my contributions to the world of the written word. Contributions which, should you care to cast an eye over them Sir Isaac, would I hope have you agreeing with me that there I can rest my case re. my appeal to you regarding fulfilment of all the conditions set down in your Third Law of Motion. I'd contend that equal and opposite more than adequately describes my reaction to the environmentally destructive actions I've found myself involved in over recent years.

Having assessed the evidence, Sir Isaac, should you still have any doubts over my commitment towards atoning for my actions, before condemning me out of hand could I at least ask for input from a renowned expert on the subject of action and reaction to be taken into account?

No one, I would argue, is better placed to contribute to such an enquiry than Ms Angularity. Not only is she fully cognisant with my performance action and reaction-wise but if there's anyone knowing more about motion, I've yet to come across them.

Although, on second thoughts, maybe the less said about that the better. That, as they say, is a whole other story.

END

Other Works

by

Mark Newham

**For sample chapters and availability of all books see:
http://moriartimedia.com/the-works**

SNOW-DODGING FOR UMPTEENAGERS

"Snuggle up by a roaring fire and treat yourself to a rare old laugh" *Saga Magazine*

Of all the bars in all the towns in all the world I walk into this one. Muted mutterings shrank to low conspiratorial whispers. Pirate eyes narrowed and glinted in dark corners. Plots for separating the fresh meat from his money suffused the bar room fug and just one thought went through my mind. Not again.

Had I learned nothing from all that time on the road? Was twenty years not enough to know that no matter how welcoming a bar/restaurant/hotel/guesthouse/resort looked on first sight, less-than welcome surprises could still await within?

Presumably not. If it had been, I'd have stopped mistaking dens of iniquity for havens of serenity, crooked paths for the straight and narrow and my search for the perfect bolthole from winter would be done with by now.

But it wasn't and that left only one option. Back gingerly out of the door, turn on my heel and resume the search elsewhere. Somewhere in the global winter sun belt there had to be a place that ticked all the snow-dodging boxes for certain age people like me and I wasn't going to stop until I found it.

Unless something stopped me first.

LIMP PIGS

"Unique... Inspiring... " *BBC*

China isn't changing. Well, it is, but not nearly as much as they'd have you believe. And unless real change comes to China soon, a long cold winter of social discontent looms. Years spent working in the gearbox of China's propaganda machine left Newham unable to conclude otherwise. Attached to two separate Chinese state news agencies between 2003 and 2008, Newham left the country convinced China is politically moribund – as authoritarian, as repressive and as unyielding as it was under Chairman Mao.

Set against China's staggering economic transformation of recent years, Newham says it's this disparity which could ultimately prove China's undoing. The country has become a child with legs growing at unequal rates. Unless something is done soon to address the political/economic inequity, ultimate imbalance is, he believes, inevitable.

Presented in the form of an irreverent memoir-with-attitude of his time working for the Xinhua News Agency and the Beijing Olympics News Service, *Limp Pigs and the Five-Ring Circus* was published in 2011 and ranked **Number One** in Amazon's censorship category for several weeks.

Revised on the inauguration of Xi Jinping as China's president in 2012, *Limp Pigs 2013* is the e-book update of the original.

COMETH THE YUAN

"Beautifully rendered... " *Guardian*

Published in 2014 and set in the not-too-distant future, *Cometh the Yuan* is a work of speculative satirical fiction envisaging the not unlikely growing extent of China's influence on the West.

Having already used its economic might to re-colonise most of the developing world, China is now eyeing more challenging targets. Chinese tendrils are already deep into western commerce and industry but that's not enough for China's ambitious leaders. Western political targets are now in China's spotlight and a campaign is launched to infiltrate western seats of power via the services of an unsuspecting multi-billionaire critic of China.

Hong Kong property magnate Harry Wong finds himself hoodwinked into participating in a takeover exercise *par excellence* courtesy of Chinese deceit and Wong's love of cricket. Inculcated into the game at Oxford University, Wong's greatest ambition is to become cricket's new supremo. The man to whom all cricket bows its head. With full Chinese support, Wong's takeover target is none other than Lord's Cricket Ground, the spiritual home of the game. China, it emerges, has confused Lord's with THE Lords – Britain's upper house of parliament.

Can Wong succeed in taking over one of Britain's national treasures? Not if the Marylebone Cricket Club can help it. Almost by accident the bungling historic guardian of Lord's finds itself at the forefront of a battle to combat China's attempt to worm its way into western politics by the back door.

PLUNDERLAND

"More gold from Mr Newham" – *Amazon review*

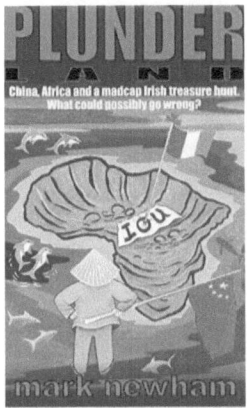

Kearney and Ryan have been in Africa so long they're becoming choc-ices in reverse.

That was never the plan. A year or two at most. Surely that'd be long enough to make the killing they know is out there. Then they'd be free. Out of Africa to enjoy the fruits of their dubious exploits at their leisure.

Thirty years on and Ireland's answer to Stanley and Livingstone are all but spent. Still no closer to unearthing that elusive pot of gold than the day they started looking both know their next hurrah will be their last.

So this time they're breaking the habit of a lifetime and doing some planning. They have something Africa's latest colonial invader China wants and, the pair are betting, will pay any price to get.

With all the bases covered, all eventualities foreseen, what could possibly go wrong?

What indeed.

**For sample chapters and availability
of all books see:
http://moriartimedia.com/the-works**

www.ingramcontent.com/pod-product-compliance
Lightning Source LLC
Chambersburg PA
CBHW031103080526
44587CB00011B/798